"The only way to save our planet and protect the climate is a global 100% renewable energy system, and solar energy will be the key component. In this book, Gregory F. Nemet traces the path of solar PV from its beginnings with impressive detail and insight to show that effective climate protection is within reach."

Hans-Josef Fell, Energy Watch Group, Germany

"*How Solar Energy Became Cheap* provides a comprehensive overview of the long pathway that PV took from a scientific idea to the world's most inexpensive form of electricity. It shows the central role China has played as well as the international linkages that have been so crucial, and it provides much needed guidance for how we can use the lessons of solar to accelerate innovation in the other climate technologies we will need."

Peng Zhou, China University of Petroleum, China

"Gregory F. Nemet has written a comprehensive and engaging treatise answering a crucial question—how did solar energy get so cheap? For decades, solar energy was a fringe energy source, but a confluence of public policies and private entrepreneurship in a few pioneer countries led solar to become the cheapest, fastest-growing energy source on Earth. Nemet's book provides a theoretically coherent explanation for this phenomenon and distils lessons that other technologies essential for combating climate change will need to follow. The book is peppered with fascinating anecdotes and backed by an impressive breadth of original research."

Varun Sivaram, Chief Technology Officer, ReNew Power, India

HOW SOLAR ENERGY BECAME CHEAP

Solar energy is a substantial global industry, one that has generated trade disputes among superpowers, threatened the solvency of large energy companies, and prompted serious reconsideration of electric utility regulation rooted in the 1930s. One of the biggest payoffs from solar's success is not just the clean inexpensive electricity it can produce, but the lessons it provides for innovation in other technologies needed to address climate change.

Despite the large literature on solar, including analyses of increasingly detailed datasets, the question as to how solar became inexpensive and why it took so long still remains unanswered. Drawing on developments in the US, Japan, Germany, Australia, and China, this book provides a truly comprehensive and international explanation for how solar has become inexpensive. Understanding the reasons for solar's success enables us to take full advantage of solar's potential. It can also teach us how to support other low-carbon technologies with analogous properties, including small modular nuclear reactors and direct air capture. However, the urgency of addressing climate change means that a key challenge in applying the solar model is in finding ways to speed up innovation. Offering suggestions and policy recommendations for accelerated innovation is another key contribution of this book.

This book will be of great interest to students and scholars of energy technology and innovation, climate change and energy analysis and policy, as well as practitioners and policymakers working in the existing and emerging energy industries.

Gregory F. Nemet is Professor at the University of Wisconsin–Madison in the La Follette School of Public Affairs, USA.

HOW SOLAR ENERGY BECAME CHEAP

A Model for Low-Carbon Innovation

Gregory F. Nemet

First published 2019
by Routledge
2 Park Square, Milton Park, Abingdon, Oxon OX14 4RN

and by Routledge
52 Vanderbilt Avenue, New York, NY 10017

Routledge is an imprint of the Taylor & Francis Group, an informa business

© 2019 Gregory F. Nemet

The right of Gregory F. Nemet to be identified as author of this work has been asserted by him in accordance with sections 77 and 78 of the Copyright, Designs and Patents Act 1988.

All rights reserved. No part of this book may be reprinted or reproduced or utilised in any form or by any electronic, mechanical, or other means, now known or hereafter invented, including photocopying and recording, or in any information storage or retrieval system, without permission in writing from the publishers.

Trademark notice: Product or corporate names may be trademarks or registered trademarks, and are used only for identification and explanation without intent to infringe.

British Library Cataloguing in Publication Data
A catalogue record for this book is available from the British Library

Library of Congress Cataloging-in-Publication Data
Names: Nemet, Gregory F., author.
Title: How solar energy became cheap : a model for low-carbon innovation / Gregory F. Nemet.
Description: Abingdon, Oxon ; New York, NY : Routledge, 2019. | Includes bibliographical references and index.
Identifiers: LCCN 2019003044| ISBN 9780367136574 (hbk) | ISBN 9780367136598 (pbk) | ISBN 9780367136604 (ebk)
Subjects: LCSH: Solar energy industries--Government policy. | Photovoltaic power generation--Cost effectiveness. | Renewable energy sources--Cost effectiveness.
Classification: LCC HD9681.A2 N46 2019 | DDC 333.792/33--dc23
LC record available at https://lccn.loc.gov/2019003044

ISBN: 978-0-367-13657-4 (hbk)
ISBN: 978-0-367-13659-8 (pbk)
ISBN: 978-0-367-13660-4 (ebk)

Typeset in Bembo
by Taylor & Francis Books

For Heidi and Pamina

CONTENTS

List of illustrations   xi
Foreword   xiii
Preface   xv
Acknowledgements   xvii
List of acronyms   xx

1 Introduction   1

2 Answer   30

PART 1
Creating a technology   53

3 Scientific origins   55

4 US technology-push   65

PART 2
Building a market   83

5 Japanese niche markets   85

6 German demand-pull   106

PART 3
Making it cheap — 131

7 Chinese entrepreneurs — 133
8 Local learning — 159

PART 4
Doing it again — 175

9 Solar as a model to follow — 177
10 Applying the model — 190
11 Accelerating innovation — 212

End matter — *223*
Index — *227*

ILLUSTRATIONS

Figures

1.1	Experts' predictions (gray diamonds) of 2030 cost of PV electricity. Actual costs shown from 1990 through 2017.	12
1.2	Contribution of PV to electricity supply.	14
1.3	Milestones in the evolution of inexpensive PV.	18
2.1	Share of world PV production.	32
2.2	Installations and production by the four leading PV actors 1992–2016.	33
2.3	Illustrative curve of a progression of markets for PV electricity. Circle indicates power purchase agreements (PPAs) signed in 2017.	38
3.1	Prior art leading to the Bell Labs patent for a Solar Energy Converting Apparatus.	60
4.1	Trends in the salience of energy as a social concern. Counts of the word "energy" in articles appearing in the *New York Times* 1970–2017.	68
4.2	US Federal PV R&D Budgets (millions of 2017$s).	72
4.3	Learning curve forecast for PV modules from 1975.	73
5.1	Japanese solar PV funding (millions of 2017$s). For comparison, the vertical scale is the same as for Figure 4.2 on US funding.	95
5.2	Prices of installed systems in Japan from 1994 to 2005.	100
5.3	Share of world PV market by application area.	101
6.1	Subsidy levels of the German feed-in tariff (FiT), the surcharge applied to rate payers, and installations in Germany.	118

6.2 Annual production (MW) by Q-Cells, Sharp, and Suntech. 122
6.3 R&D investment by PV equipment makers (millions of euros). 126
7.1 Share of global production by the largest firm in the world. 136
7.2 Price of purified silicon. 151
7.3 Average selling price of modules. 156
8.1 Share of costs for hardware (globally sourced) and soft costs (local) in the US in 2016. 160
8.2 Sources of change in the prices of installed residential PV systems in the US. 163
9.1 Comparison of development timeline for PV and direct air capture. 185
10.1 Comparison of upscaling in PV to upscaling goals for direct air capture. 193
11.1 Four innovation models for four technology types. 216

Tables

2.1 Shares of world PV manufacturing and installations. 33
4.1 US Block Buy Program 74
6.1 Global manufacturing equipment providers in the PV industry in 2018. 128
10.1 Status of current effort and future potential of innovation accelerators for direct air capture. 201
10.2 Status of current effort and future potential of innovation accelerators for small nuclear reactors. 204
10.3 Multiple models for supporting low-carbon innovation. 205

Boxes

1.1 PV terms and units. 9
2.1 Innovation terms. 34
2.2 Milestones in the history of PV. 35
2.3 Components of a photovoltaic system. 46

FOREWORD

Photovoltaic solar energy conversion is currently booming. As Greg Nemet notes in his timely new book, the electricity it provides is now inexpensive compared to almost every other electricity generation method and, even more importantly, is likely to get cheaper still. The only downside is that this outcome has taken decades to achieve. Greg provides a detailed analysis of how photovoltaics became so cheap and why it took so long. This provides a prelude to addressing the important question as to how this experience can be used to more quickly develop other urgently needed low-carbon technologies.

After the initial photovoltaic discoveries in the 19th century in both Europe and the US, Greg identifies distinct contributions in the 20th and 21st century by the US, Japan, Germany, Australia, and China that were key to the massive scale of photovoltaic production we see today. Important milestones are identified as the Bell Labs breakthrough with the first efficient silicon cells in 1954 and the major US government R&D program and procurement effort beginning in the mid-1970s. Japan and Germany followed by very successfully contributing to market development. The Japanese contribution came through the country's "million roof program," encouraging uptake through subsidies to private homeowners, with these steadily reducing as system costs dropped. This was followed by the even more impactful German "feed-in tariff" program. This not only increased the pace of market development, but also encouraged many new entrants into manufacturing. It also encouraged the development of standardized production lines provided by largely German equipment manufacturers, supporting technology diffusion, as well as rapidly evolving product quality and manufacturing scale. The Australian contribution came from training many of the entrepreneurs and engineers who successfully established manufacturing in China, initially as Australian–Chinese joint ventures. China has made the final contribution through its ability to rapidly increase the scale of the manufacturing industry, reducing costs to the incredibly low levels we see today.

Greg argues that, based on this experience, low-carbon technologies that are analogous to solar can pursue an accelerated innovation pathway to attain meaningful scale in time to positively impact climate change. Government actions can play a role in catalyzing each of the nine innovation accelerators so identified. In a nutshell, these are: continuous R&D; public procurement; trained workforce; codified knowledge; disruptive production; robust markets; knowledge spillovers; global mobility; and, finally, political economy, by which Greg means clearing the obstacles that will be deployed by large economic actors that stand to lose from the adoption of low-carbon technologies. Greg recognizes that addressing climate change is a daunting challenge, but many good things are happening and we are moving in the right direction—we just need to move more quickly.

Martin A. Green
Scientia Professor, University of New South Wales, Sydney.

PREFACE

When I began working on solar energy in 2002, the technology was seen as an intriguing novelty, serving a niche, but widely dismissed as a serious answer to social problems associated with energy use. Since then solar photovoltaics (PV) has become a substantial global industry—a truly disruptive technology that has generated trade disputes among superpowers, threatened the solvency of large energy companies, and prompted serious reconsideration of electric utility regulation rooted in the 1930s. More favorably, its continually falling costs and rapid adoption are improving air quality and facilitating climate change mitigation. PV provides some of the lowest cost electricity in the world and prices in 2018 are now below where even the most optimistic experts expected they would be in 2030.

Despite the large literature on solar we still do not have a dispositive answer to the question: *How did solar become inexpensive?* This ignorance persists because we are missing a *comprehensive assessment*, one that is truly *global*, documents the *historical evolution* of the institutional context in which solar has developed, and deals with the *full supply chain* of the industry—from sourcing silicon, the primary input material, to the activities of people installing panels on roofs, to the motivations behind adoption behavior.

As a 2017 Andrew Carnegie Fellow I have had the opportunity to dive deeply into this question, drawing on new data sets, analyses, and a growing literature. The fellowship enabled me to conduct extended interviews with approximately 70 individuals in 18 countries. I have become convinced that this mixed-methods approach adds insight that would be missed if focusing on using any single method. The quantitative analyses that colleagues and I have conducted have been valuable in distinguishing the sources of change in PV and in characterizing which of them have been most important. Qualitative data collection provides different types of insight—interviews have revealed potential causal factors that I have not seen included in regression analyses to date. For example, the tacit knowledge about aspects of PV that resides in people's heads, combined with the international mobility of those people, has been crucial in globalizing PV technology and creating new opportunities for cost reductions. The concept of National Innovation Systems provides a theoretical structure for this assessment and emphasizes that we should expect distinct national contributions to emerge from this international system, rather than thinking of it as a homogenous global knowledge stock. Quantitative and qualitative data collection provide different information about the research question, often in a complementary way. More importantly, each approach informs the other. Quantitative analysis can help establish the value of information for unknown quantities, which can then become developed into interview questions. Conversely, I anticipate that possible causal factors revealed in interviews are good future candidates to be operationalized into quantitative variables, and that can be enabled through the proliferation of quality international data and creative approaches to econometric identification.

Solar PV is exciting not just because of the massive solar resource available and low current prices, but because of how far solar has come. I am convinced that the payoff from understanding the reasons for solar's success is not just in taking full advantage of solar's potential, but in learning how to support other low-carbon technologies with analogous properties. Some technologies—those that are large and complex as well as those that are small and low-tech—do not fit the solar model. But there are a set of strong candidates, including small nuclear reactors and direct air capture, with characteristics that make them suitable for following solar's path. They can benefit from solar's drivers: scientific understanding of a phenomenon, evolving R&D foci, iterative upscaling, learning by doing, knowledge spillovers, modular scale, policy-independent niche markets, robust policy support, and delayed system integration challenges. However, it took solar six decades to become low-cost. The urgency of addressing climate change means that a key challenge in applying the solar model is to find ways to speed up innovation. A second important research question is thus not just how solar became cheap, but also *why it took so long*. The capstone of this project is a set of nine innovation accelerators—actions that would have sped the development of PV and which could be applied to new low-carbon technologies that fit the solar model. These accelerators include: 1) continuous R&D, 2) public procurement, 3) trained workforce, 4) codify knowledge, 5) disruptive production, 6) robust markets, 7) knowledge spillovers, 8) global mobility, and 9) political economy. My perspective is that committed government action in multiple jurisdictions can enhance each of these nine innovation accelerators and stimulate improvement in and adoption of the broad set of technologies we will need to address climate change.

ACKNOWLEDGEMENTS

I am grateful to many people who helped bring this project to fruition. Foremost, I want to thank the incredible research assistants at the La Follette School of Public Affairs at the University of Wisconsin-Madison who worked on this book over the past two years. Atiya Siddiqi delved deeply into the relationships between Germany and China, as well as conducting interviews with experts in Chile, Pakistan, the US, and on the ground in South Africa. Among other contributions, Travis Shoemaker figured out the roots of the Bell Labs solar battery invention and put together a more comprehensive chronology of MITI and Sharp in Japan than any I have seen in English. Rohan Rao used his data collection and analytical skills to make sense of trends in solar soft costs and the prices of purified silicon. Mikhaila Calice was tremendous in pulling together everything that needed to be in the last six months of the project. Her edits of the entire manuscript in late stages enhanced the readability substantially. I also credit the students in the Energy Analysis and Policy program for their interest, questions, and engagement over the past 12 years of my participation in that program. Multiple insights from those interactions made it into this book. The La Follette School of Public Affairs has been an excellent place to work and I appreciate Don Moynihan's support for this project at its earliest stages.

I cannot express how grateful I am to the more than 70 individuals who took the time to offer their expertise, often at considerable length, to ensure I understood their points. Much of the detail they provided I had never seen in any other sources. Their willingness to identify and make introductions to other experts made a big difference to the project.

My hosts at various institutions were essential to making this project truly global in scope. Ahmad Khaiyat at King Abdullah University of Science and Technology and at Saudi Aramco was tremendously helpful as a guide. Jiaqi Lu at the Brookings-Tsinghua Center in Beijing provided translation as well as deep historical context during my time in Beijing. Hao Ding was an excellent host for my time spent in Nanjing. Professor Masahiro Sugiyama at the University of Tokyo made my stay there highly engaging and was a huge help in introducing me to solar experts in Japan. I am grateful to Karsten Neuhoff at DIW and Jan Steckel at MCC for arranging my extended stays at those two institutions in Berlin.

Three anonymous reviewers provided careful reviews of the prospectus and initial chapters at an early stage of the project. Several individuals provided detailed sets of comments on drafts of chapters at various stages: Professor Martin Green, Dr. Osamu Kimura, Professor Frank Laird, Dr. Jan Lossen, Professor Karsten Neuhoff, Dr. Eric O'Shaughnessy, Dr. Zhengrong Shi, Dr. Minoru Shimamoto, and Professor Michele Zanolin. I appreciate their efforts, which improved the text, corrected my mistakes, and added detail, which in many cases only they knew about. Will Lamb at MCC-Berlin and Matt Wisniewski at the Wisconsin Energy Institute provided excellent graphical design contributions.

Advisors and mentors made major impressions on me. Their positive impact on my professional development has been essential for this project: Dr. Brian Arthur, Professor Adrian Bailey, Professor Dan Kammen, Professor Arnulf Grubler, Professor David Mowery, Dr. Margaret Taylor, Mason Tenaglia, and Professor David Weimer.

Other colleagues and collaborators have been a rich source of feedback, ideas, and stimulating conversations. They include, from the Global Energy Assessment: Kelly Sims Gallagher, Lena Neij, and Charlie Wilson; from my time as a visiting fellow at the Mercator Research Institute on Global Commons and Climate Change (MCC-Berlin): Felix Creuzig, Ottmar Edenhofer, Sabine Fuss, Anastasis Giannousakis, Michael Jakob, David Klenert, Jan Minx, and Jan Steckel; from my time as a Research Fellow at the German Institute for Economic Research (DIW): Karsten Neuhoff, Martina Kraus, and Vera Zipperer; and from my time as a Visiting Scholar at the Harvard Kennedy School: Laura Anadon. I also learned much from working on expert elicitations with Erin Baker, Valentina Bosetti, and Elena Verdolini; from my first project on direct air capture in 2009 with Adam Brandt; and from work on electric vehicles with Will Sierzchula, as well as on patents with Tobi Schmidt. My knowledge on this topic was enriched by working with Galen Barbose, Naim Darghouth, Ken Gillingham, Robert Margolis, Varun Rai, and Ryan Wiser over many years analyzing solar prices as part of the Academic Partners Program led by Lawrence Berkeley National Laboratory. The Energy Analysis and Policy program at University of Wisconsin-Madison attracts amazing students and I have also benefited from working with Professors Tracey Holloway, Bernie Lesieutre, Jonathan Patz, and Paul Wilson. All of these individuals have had an impact on the thinking and analysis that went into this book project.

I thank Matt Shobbrook at Routledge for taking a chance on this project early on.

The financial support I received from the Carnegie Corporation of New York made this project possible. Without the funding from their Andrew Carnegie Fellows program for such a broad, global, and interdisciplinary study, this project would have been much less ambitious. I appreciate the University of Wisconsin-

Madison, the College of Letters and Science, and particularly the La Follette School of Public Affairs for their stewardship of my Carnegie proposal.

I cannot overstate how fortunate I have been to have such a supportive family. My parents, Frank and Cindy Nemet, as well as my sister, Caroline Zanolin have always been there. My grandfather Thomas King inspired me to want to write from an early age. Melanie Mikkelsen was extremely supportive during this project. I am continually encouraged by the enthusiasm I see in my children, Heidi and Pamina. I dedicate this book to them and to their generation who have the most at stake with climate change.

LIST OF ACRONYMS

a-Si	Amorphous silicon
ADWEC	Abu Dhabi electric utility company
AIST	Japan's Agency for Industrial Science and Technology
ASE	Applied Solar Energy: Angewandte Solarenergie
BECCS	Bio-energy with carbon capture and sequestration
Bq	Becquerel
CCS	Carbon capture and sequestration
CdTe	Cadmium Telluride
CDU	German Center-right Christian Democrat party
CE	Carbon Engineering
CEC	California Energy Commission
CIGS	Copper indium gallium selenide
CPUC	California Public Utility Commission
CTO	Chief Technical Officer
DAC	Direct air capture
DACCS	Direct air carbon capture and sequestration
DM	Deutsch Mark
DOE	US Department of Energy
EEG	Erneuerbare-Energien-Gesetz, Germany's Renewable Energy Law
ERDA	Energy Research and Development Administration
ETH	Swiss Federal Institute of Technology
ETIS	Energy technology innovation system
ETL	Electrotechnical Laboratory
EU	European Union
EVA	Ethylene vinyl acetate
EW	Energiewende, Germany's 40-year renewable plan
FiT	Feed-in tariff

FOAK	"First-of-a-kind"
Fraunhofer-ISE	Fraunhofer Institute for Solar Energy Systems
FSA	Flat Plate Array Solar Array Project
GPTs	General purpose technologies
GSF	Global Solar Fund
GT	Global Thermostat
GW	Gigawatt
IEA	International Energy Agency
IPOs	Initial Public Offerings
ISO4	Interim Standard Offer Contract #4
IRENA	International Renewable Energy Agency
JPEA	Japan Photovoltaic Energy Association
JPL	NASA Jet Propulsion Lab
JSEC	Japan Solar Energy Company
kW	Kilowatt
kWh	Kilowatt hour
LbD	Learning by doing
LCFS	Low-carbon fuel standard
MEMC	Monsanto Electronic Materials Company
METI	Japanese Ministry of Economy, Trade, and Industry
MITI	Japanese Ministry for International Trade and Industry
MSART	MW System Integrated Modular Advanced Reactor
MW	Megawatt
MWh	Megawatt-hour
NDRC	National Development and Reform Commission
NEDO	New Energy Development Organization
NETs	Negative emissions technologies
NFP	National Fabricated Products
NGOs	Non-governmental organizations
NIS	National Innovation System
NRC	US Nuclear Regulatory Commission
NREL	National Renewable Energy Laboratory
NSF	National Science Foundation
PACE	Property Assessed Clean Energy
PCAST	President's Council for Advisors on Science and Technology
PECVD	Plasma enhanced chemical vapor deposition
PPA	Power purchase agreement
PTC	Production Tax Credit
PURPA	Public Utilities Regulatory Policy Act
PV	Photovoltaics
PVSEC	Japanese PV Science and Engineering Conference
PVTEC	Power Generation Technology Research Association
R&D	Research and development
RANN	NSF Research Applied to National Needs program

RPS	Renewable portfolio standard
SAMICS	Solar Array Manufacturing Industry Costing Standards
SERI	Solar Energy Research Institute
SMR	Small modular nuclear reactor
SNR	Small nuclear reactor
SOE	Chinese state-owned enterprise
STE	Solar thermal electricity
STI	Solar Technology International
StrEG	Stromeinspeisungsgesetz, Electricity Feed-in Law of 1990 in Germany
TI	Texas Instruments
TPO	Third-party ownership
TWR	Traveling Wave Reactor
UNSW	University of New South Wales
W	Watt
WIN	Western Initiative for Nuclear program
WTP	Willingness-to-pay
x-Si	Crystallized silicon

1
INTRODUCTION

We desperately need new ways to accelerate technology development and its adoption in order to address energy related challenges such as climate change. We need models that show what it takes to be successful. My approach is to look for successful outcomes that might provide models for early stage technologies to follow. There are many candidates for case studies exemplifying elements of success: nuclear power made the largest new contribution to electricity supply; vehicle fuel economy provides an example of rapid policy-driven adoption; hydraulic fracturing illustrates rapid market-driven adoption; pollution controls show that actors will adopt technologies even when their benefits are purely external; and smart grid technology is novel in that consumers interact with the energy system more directly. In searching for an appropriately successful case to study, I prioritize learning—new knowledge that improves technology—because that will make the case's lessons most applicable to nascent technologies that still need to develop further. I thus select the energy technology that has improved the most, solar photovoltaics (PV). PV represents the technology effort that involved not the largest investment but the most dramatic technological outcome, the most learning. It is the technology that most clearly shows what support for innovation can achieve.

The costs of solar PV modules have fallen by more than a factor of 10,000 since they were first commercialized and are now cheaper in sunny places than any other form of electricity. Since its first application—on a satellite—in 1957, PV has always found customers who value it. PV has served a sequence of increasingly large niche markets with decreasing willingness to pay. This niche market strategy worked in part because the technology could function at any scale—the largest PV applications are a billion times larger than the smallest. Entrepreneurs were crucial throughout PV's history, from California in the mid-1950s to China in the 2000s. Public institutions played critical roles—especially funding R&D and subsidizing early markets. The German feed-in tariff of 2000 was instrumental in enabling the industry to achieve scale, and catalyzing investment in installation, production facilities, and PV-specific manufacturing equipment. Five key countries—the US, Australia, Japan, Germany, and China—made distinct contributions that emerged from each country's national innovation system. Each built on the work of previous leader countries with global flows of knowledge, in people, devices, and machines, catalyzing a truly global innovation system. However, all of this occurred too slowly.

Motivation

My motivation for writing this book is that even though PV's potential as a climate change mitigation tool is vast, it can contribute even more to climate stability by serving as a generalizable model for other low-carbon technologies. If we decide that PV is a useful model to follow, having the broad outlines of how PV worked is insufficient to apply to other technologies. We need a detailed understanding, we need to include variables that may be omitted in statistically-based regression analysis, and we need to make all of it happen faster because PV took far too long for its path to be directly applicable to other technologies we will need to address climate change.

Before proceeding with an overview of the case of PV, I first briefly outline two premises that support this motivation. First, climate policy competes with other social objectives, even within the energy policymaking domain. Climate stability is only one of many societal benefits of an improved energy system. Second, the idiosyncrasies of the climate problem itself make its innovation needs distinct. We need to account for public goods, a role for government, long time horizons, and consequently, urgency in technological development.

Multiple objectives

Despite its efforts, social science is missing the inviolable laws that physical sciences use to build a foundation of understanding complex phenomena. For example, social science does not have rules comparable to the generality of gravity or the first and second laws of thermodynamics. When patterns do emerge, hyperbole is typical, and social scientists arrive at "iron laws" (Pielke, 2010). If there is an iron law of energy policy I suggest it is this: *energy policymaking always involves multiple objectives.*

A foundational premise that I take in understanding approaches to addressing energy related challenges is that energy policymaking inherently involves reconciling multiple objectives. As a society, we want energy that is cheap, reliable, and clean. Cheap energy means it is affordable, but also that the economy avoids macro-economic shocks as experienced in the 1970s. For about a third of the world's population, cheap means moving up the energy ladder to access modern energy carriers; away from traditional biomass and toward electricity. Reliable energy is about having a secure source of supply. Much of that emphasis is rooted in past military conflicts, in which access to fuel was decisive. Reliability implies domestic supply sources, or at least friendly international ones. It is not just about the prices paid but also the ways in which energy security affects the ability of actors to negotiate on areas outside of energy. Clean is multifaceted as well. It includes avoiding the human health and ecosystem effects of air pollution, as well as the damages from an unstable climate.

Billions of people have a stake in the outcome and they do not agree on whether to prioritize affordability, reliability, or cleanliness. This disagreement matters because substantial tradeoffs exist among the alternatives. Fossil-derived synthetic fuels can give us reliability but are not cheap or clean. Nuclear power is clean but is not cheap. Coal is cheap in many places but is not clean. These objectives have shifted over time. Concerns about resource depletion in the early 1970s, as well as about pollution, generated nascent activity in US energy policymaking. With the 1973 Arab Oil Embargo, the focus shifted to energy security and affordability. Much progress has been made but those challenges still remain. In climate change, over the past 30 years we have added an even more demanding energy-related challenge.

One approach to these competing priorities is to do the best we can with these tradeoffs and make a compromise among the goals of cheap, clean, and reliable energy. Deliberative democracy provides one avenue to negotiate a solution (Ryan et al., 2014). But because people's preferences can shift over time, we often see instability in policy and even sudden shifts (Nemet et al., 2014). The energy system is inherently slow to change, and the climate system even more so. We thus need approaches that are persistent (Nemet et al., 2016) otherwise, the tradeoffs will recur in a cycle, as one insufficiently addressed objective demands attention, for example during a crisis, while others lose salience setting up conditions for the next crisis (Downs, 1972). Another approach is to change the choices we have available to us so that the cheap, clean, and reliable goals do not conflict. Changing the choices we face requires innovation. For many, that approach is unrealistic because it evades the harsh reality of tradeoffs and making difficult choices among alternatives. It's been referred to as "magical" thinking, of betting that things that do not exist today will exist in the future (Klein, 2015). But this "magic" is exactly the promise of innovation. It is the magic that has transformed the world from a Hobbesian competition for scarce resources to one in which ideas, and crucially their application, can help generate the goods and services that society wants with fewer resources involved in producing them. The continuous reduction in energy input per output of GDP provides the most comprehensive evidence of the effects of innovation on resource use (Ausubel and Waggoner, 2008). Energy innovation is powerful because it can alleviate these tradeoffs and thus provide solutions that are more persistent than democratically achieved compromises.

The decarbonization challenge

While innovation can make reconciling conflicting energy objectives less contentious, for climate change innovation isn't just helpful, it is essential. Avoiding a substantial portion of the future damages from climate change—while affordably meeting the world's growing need for energy services—will require a fundamental transformation of the means by which energy is produced and used (Nemet, 2013). Addressing climate change in a material way requires deep and broad innovation. While the focus of climate policy is typically on emissions or temperature targets, the scale of the transformation required elevates incentives for innovation to a central concern of climate policy itself. Decisions involving energy technology policy, and more specifically, policies intended to accelerate the development and the deployment of low-carbon energy technologies, lie at the center of climate policy debates.

Addressing other energy concerns—for example, affordability, reliability, and air pollution—have also involved innovation (Taylor et al., 2003). But they have tended not to require fundamental changes to the technological and regulatory systems that provide energy services. Electricity became affordable in developed countries in the 20th century through economies of scale in power plants and increased load factors achieved through large interconnected electric grids. Reliability improved by reducing demand through higher fuel economy in vehicles and expanding supply through offshore drilling and hydraulic fracturing. Air pollution improved by switching to low sulfur coal and installing pollution controls on existing power plants. In each case, the energy system was modified incrementally, rather than radically transformed. For example, most of the changes did not require changes to the supporting infrastructure.

Addressing climate change is different from these other energy problems because it will involve deep changes to the way energy is used and produced. The more ambitious targets discussed in policy debates and then adopted, such as the Paris Agreement (UNFCCC, 2015), imply rates of technological change that are well beyond historical precedents (Nemet, 2013) and will likely require non-incremental technological improvements (Rogelj, 2017). Such radical innovations may require new forms of public incentives as they often require new infrastructures. Further, it may be more difficult for private actors to capture the value in them due to their technological radicalness and tendency to provide public goods in the form of knowledge spillovers.

Energy technologies are also different from other technologies. The existing systems have been in place a long time because the equipment lasts a long time (Knapp, 1999). Contrast this longevity to smart phones, which get replaced and upgraded on a 2-year cycle—or prescription drugs that get refilled, with an opportunity to be switched, every 30 or 90 days. Power plants, transmission lines, pipeline, buildings last decades. Because of the up-front costs, energy technologies tend to have long payoff times and require many complementary, slower innovations.

Scale and urgency

Two aspects of the climate problem—scale and urgency—have profound implications for efforts to innovate to address climate change. First, the scale of the problem requires that we fundamentally transform a system that serves the entire world. The global energy system is expanding to serve new demands of people emerging from subsistence and industrializing to live modern lifestyles. Further, when dealing with climate, gigatons—billions of tons—are what matter. While solutions that have smaller impacts and do not scale to large ones may be helpful and are increasingly available, they struggle to justify attention and investment in them; we need to think in solutions at the scale of billions of tons, gigatons, of CO_2 (Herzog, 2011).

Second, despite the inertia discussed above, there is urgency to the climate problem. Certainly, inertia exists, both with energy systems and with the climate. That resistance to change is what necessitates urgency. The world first got serious about climate change in 1988. Three decades later and one would have to squint to see a bend in the curve on global greenhouse gas emissions (Jackson et al., 2017). As another perspective, consider these gigatons as a rate of change. Emissions need to decline (decarbonize) at about 5% per year for the next several decades (Millar et al., 2017). That is a far larger decline than what we have seen over the past three decades, during which emissions have continued to rise (Peters et al., 2017). We can look at previous examples. Over the past three decades, six countries have decarbonized their economies at rates faster than 5% sustained over 10 years or more (Nemet, 2013). These include China's modernization in the 1980s; Russia, Poland, and Slovakia shutting down inefficient facilities post-communism; and Sweden and France adopting nuclear power in the 1980s. To be fair, these changes had little to do with addressing climate change. If we look just at the data after 1997 when the Kyoto Protocol was signed, only the eastern European countries decarbonized at rates above 5% and only one of these countries (Russia) ratified the protocol and that was at the very end of the data period. Achieving the decarbonization rates implied by recent targets is a formidable challenge that is on the outer fringe of historical precedent. There have been some successes in relatively substantial decarbonization, but the more impressive ones were done without climate change in mind: nuclear in France and Sweden, post-soviet de-industrialization in eastern Europe, poverty alleviation in China, and hydraulic fracturing in the US. The biggest consolation one can take from these data is that we've never focused on innovation specifically for decarbonization; so presumably, with deliberate intention we could do better (Geels et al., 2017).

Many technologies needed

In part because of the scale of the change required, a broad array of technologies will need to be effective, affordable, and deployed at gigaton scale. There is no silver bullet; no one technology will do the job, not even a couple. There are several reasons to expect and want a broad set of technologies to address climate change rather than a small number of very promising ones. First, we don't know what will happen as new technologies develop and become adopted at increasing levels. This uncertainty implies we are better off with a broad portfolio of ways to address climate change (Anadon et al., 2017). Second, the world is a heterogeneous place. People want different services, delivered in different ways, from their energy system (GEA, 2012). A technology that is scalable, clean, and affordable in one place might not work somewhere else. Third, it is clear that technologies, even beneficial ones, begin to impose adverse social impacts at very large scale (Grubler, 1998). We can avoid those impacts with diversity. One implication from a policy perspective is that it is undesirable to champion or concentrate on a single technological pathway. Our policies should reflect diversity as a means to support the larger social goals of energy innovation. We won't be able to address climate change by concentrating resources on getting only a few key technologies to scale.

The government role

Public policy will be needed because individuals and companies are unlikely to undertake this effort on their own, even if their decisions will be central to the outcomes. Two justifications for the role of policy in energy innovation include the concept of market failures and a system of failure argument. First, there are at least two well-established market failures involved in climate related innovation. The agents who engage in activities that emit fossil fuels are not paying for the climate damage they cause (Tybout, 1972). Governments can correct this market failure by making those agents pay for those damages with a variety of policy instruments, such as a carbon price. A second market failure arises because nascent technologies, especially novel ones, can be copied (Teece, 1986; Noailly and Shestalova, 2017). This market failure exists because knowledge generated by investments in innovation is difficult to appropriate: it spills over from one firm to another, and similarly from one country to another. These knowledge spillovers reduce the incentives to invest in technologies, giving rise to a second public-goods problem: firms, and in some cases countries, have incentives to free ride on the technology investments of others. The problem with this innovation-related market failure is that spillovers of knowledge will lead firms to pursue too little investment in innovation (Rosenberg, 1990).

A second set of reasons comes from work on "innovation systems," in which governments need to address system failures (Carlsson and Stankiewicz, 1991; Bergek et al., 2008). In justifying government interventions one seeks to identify failures in components of the system, such as inadequately performing functions by entrepreneurs (Bleda and del Rio, 2013). New technologies competing with long established ones typically face these problems at various stages in their development.

Certainly, companies and individuals will have to make major innovation efforts to decarbonize the world economy. But the point here is that the current incentives to do so are insufficient without governments finding a way to account for negative pollution externalities, positive knowledge externalities, and innovation system failures.

Need for innovation

A wide array of groups has made the case that we need a renewed emphasis on energy innovation in order to address energy and climate concerns. This effort started with the President's Council of Advisors on Science and Technology (PCAST) studies in the 1990s, which emphasized the need for public R&D

(PCAST, 1997; PCAST, 1999; Holdren and Baldwin, 2001). Other work justified R&D investments by quantifying in dollar terms the multiple benefits of energy technologies, e.g. in terms of avoided national defense costs (Schock et al., 1999). Another set of papers emphasized the need for public R&D by documenting the declining investment in private R&D (Dooley, 1998; Margolis and Kammen, 1999; Nemet and Kammen, 2007). Data collected on R&D for developing countries highlighted the rising importance of innovation in those places (Gallagher et al., 2011). A large set of studies followed up on the US National Research Council report that recommended using expert elicitations—structured interviews to ask about expected technology outcomes—to allocate R&D (NRC, 2007). These studies used expert judgment to quantify the effects of R&D spending on a wide array of energy technologies (Anadon et al., 2017; Verdolini et al., 2018). Another set of studies looked at innovation more broadly, beyond R&D, focusing on innovation systems (Hekkert et al., 2007), the specifics of the energy technology innovation system (Gallagher et al., 2012), and a large set of case studies to develop a theory of what seems to work (Grubler and Wilson, 2014). This latter set of studies emphasizes the need to think systemically.

Reasons for optimism

The combination of massive change needed, in a short amount of time, with little substantive indication of progress after 30 years of effort, has led many observers to adopt a discouraged outlook in addressing climate change. The case supporting the discouraged perspective is plentiful, based on hard data with historical evidence. Deviations from this view are dismissed as too small. Progress in international negotiations is "just talk." It has become much easier to be taken seriously as a skeptic when addressing climate change than as a proponent of action. To be clear these are not climate change skeptics who doubt that there is a problem, but rather action-skeptics who acknowledge there is a problem worth addressing but are convinced it is just too difficult to do. Earlier in my academic career I was one of these action-skeptics. My focus in teaching energy analysis and global governance was to make clear how big the challenge of changing the energy system would be—and how difficult international cooperation would be. I frequently used the word unprecedented. With quantitative problem sets, ample readings, and my comments on their written memos and term papers, I succeeded in convincing my students how hard the problem is.

One outcome of the production of action-skeptics is a proliferation of discouragement. One sees it in the media and in the academic literature. And most concerning and disheartening to me, I have seen it in the students I teach—those who have chosen to take a class on energy systems or environmental governance. It has led me to ask why I bother to work on this problem—and why so many others do. I think people work on climate change because first, they think it is worthwhile to try and second, they are optimistic that there really is a way out. Following the UN Copenhagen Summit on climate change in December 2009, seeing the intense discouragement among students, I decided to put together a list of reasons to be optimistic about dealing with climate change. These include: emerging collective action, learning from policy experience, and precedents in other areas. I discuss the full list in Chapter 11.

Over the past few years, one item on the list became by far the easiest to justify, the most convincing. The one that actually seemed to perk jaded students from their slumped shoulders and distracted looks was "improvements in technology." In many cases it bolsters the credibility of the other reasons. That is also where my teaching most directly intersects with my research. I work on innovation in energy systems. If we want to proactively stimulate innovation, we need successful models. We also need to understand why they were successful. The one technology that has changed more than any other energy technology is PV.

Photovoltaics are a success

In searching for successful models of energy innovation to address climate change, one should look first at PV. The cost and performance of PV have improved more than any other energy technology (McDonald and Schrattenholzer, 2001; Rubin et al., 2015). It is now astonishingly inexpensive. Not only is PV cheap compared to what it cost only a few years ago, but it is also cheap compared to almost every other electricity generation method. It is likely to get cheaper still. Cheap PV has been an amazing achievement by human civilization, but it was never inevitable. There have been several false dawns over the past 70 years of PV's evolution. Note that throughout this book, I reflect the terminology of the broader literature and use solar, PV, and solar PV interchangeably. While it is true that the term solar can encompass more than PV, I have found in teaching interdisciplinary college courses that using acronyms makes material less accessible and that regularly returning to full words can be helpful even if sacrificing concision.

PV Changed the most

The cost of the core PV technology, a module, has fallen dramatically. PV's progress is often compared to the improvements in computer processors (Farmer and Lafond, 2016; Hutchby, 2014). A megawatt-hour (MWh) of PV electricity in 1957, for its first commercial use, cost about $300,000 in today's dollars. A megawatt-hour is a month's worth of electricity for the average US household. Today in a sunny location, that megawatt hour costs only $20. Thus, the cost of that electricity has fallen by four orders of magnitude, a factor of 15,000 cost reduction.

One consequence of the tremendous progress in module costs, is that "soft" costs now dominate the total cost of PV electricity (Barbose et al., 2017). These soft components involve activities with installing: roofers and electricians; marketing a novel technology; sales, often on a one-roof-at-a-time rate; financing; new business models; refreshing electric utility regulation; policy innovation; and crucially, integration into the surrounding technological system. But "soft" costs are improving too (Gillingham et al., 2016) and are catching up to the incredible progress seen in module costs (O'Shaughnessy, 2018).

PV is cheap today

PV is not just a promise for the future, it is stunningly cheap today. In many places, it is the least expensive way to produce electricity (Lazard, 2018). PV today in sunny locations is likely the cheapest way humans have ever found to generate large-scale electricity. In October 2017, Masdar (Abu Dhabi) and Électricité de France submitted a bid to supply power to the Saudi Arabia power grid at $17.9 per MWh. Saudi Arabia's ACWA Power had the second lowest bid at $22 per MWh and eventually won the contract due to its perceived higher reliability. A combination of factors made these bids so low, including an extremely favorable solar resource, as well as the falling cost of PV hardware. Easy connections to the transmission system, free land, and low financing costs also contributed (Dipaola, 2017).

What is most impressive about the recent costs of PV power projects is not just the record in Saudi Arabia, but the multiple contracts being signed in very different places, all at extremely low prices (IRENA, 2018). A first notable event was the claim in 2014 of US PV projects costing $50 per MWh. That price was so dramatically low that it spurred analysis to see if it was deceptively low. Using data on costs and assumptions on financing and maintenance, the calculations revealed that it was, in fact, real (Bolinger and Seel, 2015). Perhaps the biggest focusing event for people's awareness of large-scale inexpensive PV came in September 2016 when the Abu Dhabi electric utility (ADWEC) signed a contract to buy solar electricity at $30 per MWh (IRENA, 2017). That rate was below the lowest power contracts in the US of any type, including that of natural gas and coal; in fact, it would undercut nearly any price around the world. Since then, power companies have signed agreements to buy PV electricity at $29 per MWh in Chile, $27 in Mexico, $23 in Nevada, US, and in October 2017, the $22 in Saudi Arabia. In late-November 2017, Mexico signed a contract at a similar price as the Saudis.

BOX 1.1 PV TERMS AND UNITS.

KWh: A kilowatt-hour of electricity is using 1000 Watts of power for 1 hour. Electricity is typically billed in these units.

MWh: A megawatt-hour is 1000 kilowatt-hours. An average residence in the US consumes about 1 MWh in a month.

Retail electricity prices: Prices paid by end-users. For example, individual households or businesses pay retail electricity rates. Residential PV competes with retail electricity prices.

Wholesale electricity prices: Prices paid by upstream buyers of electricity which they sell to retail customers. Utility scale solar competes with wholesale electricity prices.

> **Power Purchase Agreement (PPA):** Contracts to buy power on the wholesale market are typically done as a PPA in which one party agrees to buy power for a certain period at a specified rate per MWh.
> **Constant dollars:** To make the purchasing power of a dollar comparable across time, one can extract price inflation and put units in the dollars of a chosen year, for example 2016 dollars.
> **Installed cost:** The full price paid for a PV system including hardware and connections to the electric system.
> **Hardware costs:** Costs of PV modules and inverters.
> **Soft costs:** Installed cost minus hardware costs. Typically involves labor costs for installation, as well as sales, marketing, and administrative costs.

Not just in sunny places

States in the Persian Gulf have some of the best solar resources in the world, with ample land that is not in productive use. Their primary sources of domestic energy supply are oil and gas, which find much higher prices on export markets than in domestic consumption as electricity, which is heavily subsidized. PV in the Gulf is in many ways a no brainer. However, we are now seeing inexpensive PV in less favorable places. Germany has been at the forefront of PV development for four decades, but its high willingness to pay to fund PV deployment does not by itself make PV cheap. In fact, Germany has only about one-third the sunlight of the Gulf states, or about the same as the least sunny parts of the lower-48 states in the US. Consequently, a system in Germany would produce electricity at three times the $ per MWh cost of that same system in the Persian Gulf, due to fewer MWh produced from the same equipment. In 2014, Germany converted its PV subsidy scheme to an auction so that the lowest bid in euros per MWh wins the procurement contract. Even relatively dim Germany is seeing very low prices. In 2017, the winning PV bid was at $45 per MWh. In Germany that price is competitive with coal and beats natural gas. Colorado is more than twice as sunny as Germany, but not quite at the level of the Persian Gulf. In a 2018 request for proposals, solar was bid at $30 per MWh, and a solar with battery storage bid was at $36 per MWh. Recent power purchase agreements (PPAs) for natural gas generation, which is far cheaper in the US than in other parts of the world, are in the range of $20–30 per MWh (Wiser and Bolinger, 2017).

Not just in wealthy places

Solar is not just cheap in wealthy developed countries either. India's per capita GDP (PPP) is 11% that of the US and less than half of China's. Yet India has installed more than a gigawatt of solar in each of the last four years. It is the third largest solar market after China and the US. Egypt is the latest country to install more than a gigawatt of solar. Tanzania has an active solar sector with venture backed Off-grid Electric recently raising $50 million to fund projects in Tanzania and Rwanda. Chile has one of the least expensive PV contracts in the world, amid a strong commitment to take advantage of its substantial solar resource. Faced with an energy shortage and high prices, 8% of Chile's energy matrix is now solar. South Africa also has an active sector with solar used to power off-grid water pumping, a strong niche market in rural areas. There is water in the mountains but traditional hand-pumps used to get that water break easily and are time consuming. Using solar energy reduces cost and increases efficiency. Liberia has an active solar sector as well that started with the introduction of solar powered lamps. The market is currently expanding into solar powered pumps for irrigation and rooftop PV, replacing the gas generator commonly used to electrify homes in rural towns.

Many don't get it

The size of Earth's solar resource is extremely large (IPCC, 2011). A few minutes of sunshine could power the world for a year. A square of panels a hundred miles on a side could power civilization continuously. I would guess that almost everyone with some professional involvement in the energy sector over the past 60 years has heard a version of these claims. In the past these claims were easily dismissed with a "yes, but" because solar was too expensive. My experience is that both the general population, as well as many in the energy sector, do not understand how cheap solar has become. One still hears comments like: "It is still too expensive" and "but it's not a viable replacement for baseload fossil fuel generation." Most people do not realize how tantalizingly close we are to taking advantage of a substantial part of the immense solar resource described above. In many places solar is the cheapest way to make electricity, period. Yet, people still seem to think solar is prohibitively expensive. Perhaps it takes a long time for people's perceptions to change. And perhaps the promises made—particularly in 1954, in 1979, and in 1993—created disappointment and eventually dismissal of the feasibility of solar. Many people approach solar dubiously even today.

Experts were wrong

An intriguing aspect of this solar skepticism is that even those most knowledgeable about solar technology have been overly pessimistic about it. Between 2008 and 2011, colleagues and I conducted structured interviews and surveys with 65 solar experts to elicit their expectations of solar prices in 2030 (Verdolini et al., 2015; Curtright et al., 2008; Baker et al., 2009; Anadon et al., 2011; Bosetti et al., 2012; Davis and Fries, 2011). We chose subjects to interview not as part of a random sample, but rather, as specifically targeted individuals who were among the most knowledgeable people in solar. At the time of the interviews, 2030 seemed like a date when solar might become competitive on a large scale. We compiled those results to make them consistent (Verdolini et al., 2015), and in comparing the results to reality a few observations emerge (Figure 1.1):

12 Introduction

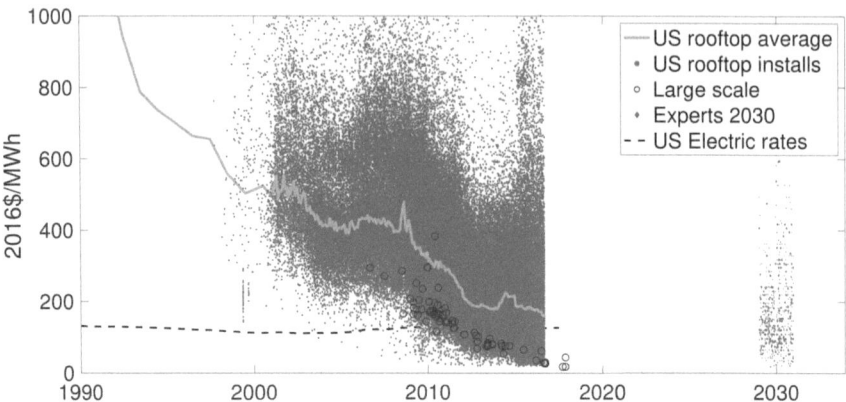

FIGURE 1.1 Experts' predictions (gray diamonds) of 2030 cost of PV electricity. Actual costs shown from 1990 through 2017.

- Actual prices being paid for solar electricity in 2017–18 are below the experts' median predictions for 2030;
- The lowest PPA contracts in 2017 described above are below the median of each the expert's most optimistic case for 2030.
- In fact, the Saudi and Mexican $18 per MWh in 2017 are below all of the experts' most optimistic predictions—again, for 2030.

Solar's progress between 2010 and today has been stunning, even to those most informed about the technology.

Models were wrong

Computer models that simulate the dynamics of the global energy system have also been repeatedly wrong about PV. Energy forecasts are notorious for being wrong anyway (Craig et al., 2002) and have been characterized as "computerized fairy tales" (Smil, 2008). When colleagues and I looked closely at PV forecasts (Creutzig et al., 2017), it became clear that the poor record in accuracy is not just random error—over-predicting PV adoption one year, under-predicting the next. Rather forecasts are consistently biased, always under-predicting PV adoption. Even the most optimistic forecast we could find of PV, from Greenpeace, under-predicted how much PV there was in reality (Greenpeace, 2015). Some forecasts were off by an order of magnitude. Why were they so wrong? In that study (Creutzig et al., 2017), we identified three reasons that models under predicted solar: missing learning by doing, missing policies, and missing cost increases in competing technologies. First, while these models have improved their characterization of technological change, none have reliably characterized the positive feedback between deployment and cost reductions, sometimes referred to as learning by doing. This interaction is at the core of solar's evolution.

Second, these models have not included direct policy support for PV in their optimization of energy supply. As I document later in this book, subsidies for PV have stimulated the market tremendously, first in Japan, then in a much bigger way in Germany, and since then in many countries, notably China. PV adoption has grown tremendously in part because of these policies. Behind the policies is strong public support which has made the policies persistent and often stringent. Most important, these policies have catalyzed the positive feedback mechanisms of learning by doing described above.

Third, models have been too optimistic about the costs of other low-carbon competitors to PV, notably nuclear and carbon capture and sequestration (CCS). These technologies have turned out to be far more expensive than anticipated. The proliferation of inexpensive natural gas has had the opposite effect in the US; it has been cheaper than expected. But so far that dynamic has been limited to the US. And even there, we are now seeing PV close to, if not already, beating natural gas (Lazard, 2018).

Churlish critics

Finally, one must acknowledge that there are those who do not want PV to become widely adopted. A variety of motivations lie behind this stance. Some seem genuinely concerned about threats to the stability of the power grid. Some are pre-occupied with the negative externalities that PV users impose on the rest of the power grid. Some still associate PV with hippies, liberals, and heavy-handed government. For some, solar panels are unfamiliar, an eyesore even. For others, PV threatens their existing business models. They point to the dire straits of the four large, German electric utilities found themselves in once PV took off there and began depressing power prices (Morris and Jungjohann, 2016).

The prevalence and strident tone of PV critics is perhaps the strongest indication that it has arrived as a serious energy technology. There are certainly concerns and limitations to existing PV systems. These include its intermittence, existing utility business models in developed countries, and financing constraints in developing countries. Some have pointed to toxics (Fthenakis and Moskowitz, 2000), energy input required (Hertwich et al., 2015), recycling needs (Davidsson and Höök, 2017), and even warming due to reducing the Earth's albedo (Nemet, 2009b). Some claims have been debunked, some need more analysis, and some need more work to overcome. But at some point, to criticize PV, to dismiss it, seems churlish. Who could possibly not acknowledge that cheap PV now provides another option to alleviate societal problems with energy production and use? PV need not be 100% of electricity or 50% of energy to be a huge asset to society's effort to transform the energy system.

A long way to go

Despite the incredible progress in PV, this is no time for solar triumphalism. PV is still only 1–2% of global electricity supply and less than 1% of total final energy. In some places such as Japan and Germany it is several times that (Figure 1.2). In others, such as US, China, and India, it is less than 1% of electricity.

Major changes are needed to adapt energy systems in order to ensure that PV is a central, rather than peripheral, component of energy systems. The complete story of PV will ultimately be written in the next couple of decades as we see whether it goes to 30–50% of electricity or stays around 1% (Sivaram, 2018b). Modernizing the system is the core challenge, both in terms of institutions and infrastructure, as well as the interaction between the two. Some places seem better prepared to manage this transition than others. But that 1% vs 50% is now a contentious debate only arises because solar PV is now the cheapest form of electricity. But this evolution has taken an excruciatingly long time. As Figure 1.2 shows, it took 70 years, longer if one looks back to the formative phase in the 19th century. PV only serves as a useful model if there is a means to speed up that model.

Research questions

The success of PV motivates the three research questions for this project:

1. How did PV become so cheap?
2. Why did it take so long?
3. How can PV serve as a model for other low-carbon technologies?

Despite the large and growing literature on solar, including analysis of increasingly detailed datasets, we still do not have satisfactory answers to the questions above. The motivation behind this book is that this ignorance persists because we are missing a comprehensive and systemic assessment of solar. We see foci on important aspects, including technical improvements (Green et al., 2016), consumer adoption behavior (Rai and Sigrin, 2013; Graziano and Gillingham, 2015),

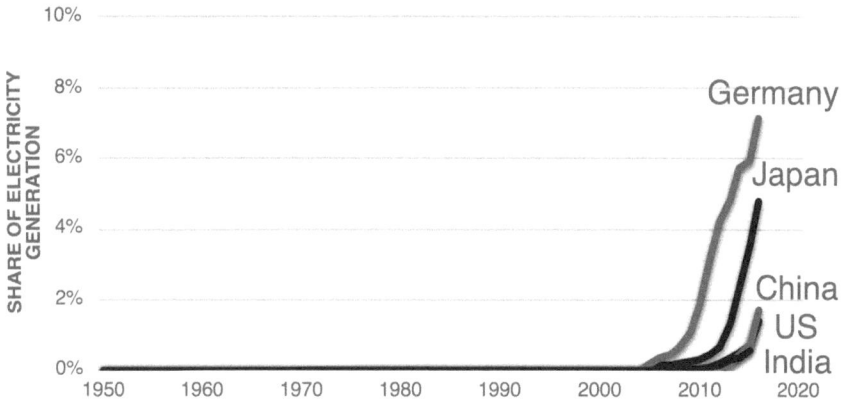

FIGURE 1.2 Contribution of PV to electricity supply.

electric grid integration (Deetjen et al., 2016; Gowrisankaran et al., 2016), and the policies of individual countries (Hoppmann et al., 2014). However, these studies provide only glimpses of insight. Really answering the question requires an assessment that is truly global (Zhang and Gallagher, 2016), documents the historical evolution of the institutional context in which solar has developed (Laird, 2001), and deals with the full supply chain of the industry—from sourcing the primary input material, silicon (Pillai, 2015), to the activities of people installing panels on roofs (Friedman et al., 2013), to the motivations behind adoption behavior (Rai and Robinson, 2013).

Further, some of the key innovators—including Martin Green in Australia, Paul Maycock in the US, Tokuji Hayakawa in Japan, Zhengrong Shi in China, as well as Hans-Josef Fell and Hermann Scheer in Germany—have had quite distinct personal motivations for pursuing cheap efficient solar energy over several decades. There is something about solar that motivates people beyond their financial self-interest (Wolfe, 2018). Those intrinsic motivations, and the career trajectories they influenced, form an essential aspect of how solar became cheap. They have been important because solar's history of commercialization has been a wild ride of multiple booms and busts going back to the 1950s with early observations back to 1839. Over 70 personal interviews have thus been the central research strategy for this project. Even though the focus is on the four key countries—US, Japan, Germany, and China—I conducted interviews with people in over a dozen countries in total because the drivers of progress, as well as the outcomes, have extended beyond those four.

Answering the three research questions in this book involves using existing data, assembling new data, and making use of the analysis conducted by myself and others. Making use of and pointing to the work of others is central because this study is being conducted by a single author. A central challenge of interdisciplinarity is that one person cannot have the deep expertise that any individual specialists might, as one can only have breadth and depth in one or two areas. The approach is therefore integrative, interdisciplinary, and comprehensive—at the risk of being superficial relative to the excellent other studies, particularly of PV's early development (Wolfe, 2018; Perlin, 1999; Laird, 2001). Given the focus on innovation in an international context, I use national innovation systems (NIS) as a theoretical framework for this book. I also make use of the energy specific work on innovation and the energy technology innovation systems (ETIS) framework.

National innovation systems

National innovation systems theory became prominent in the 1990s as a way to explain innovation in an increasingly globalized economy (Nelson, 1993; Freeman, 1995). The core concept—that countries pursue innovation in distinct ways—dates back to the work of Friedrich List, who saw similar idiosyncrasies in the way that economic actors engage with the state (List, 1841). The central insight is that, despite the steady increases in international trade, communication, and transportation, we miss important aspects of innovation by looking at it globally as emerging from a single global system. Rather, the country level is a more useful container for geographically bounding analysis of activities. The networks of relationships among firms and individuals are central to how innovation occurs; those relationships are often much stronger within a country rather than between countries. Further, the national container is useful because innovation is conditioned by distinct institutions within them: education systems, industrial relations with the state, science and technology institutions, government policies, cultural traditions, and other national institutions (Lundvall, 1992). These national characteristics condition micro-level behavior, that is how firms and individuals interact, to produce a distinct innovation system. There is a strong emphasis in this literature on looking at knowledge, knowledge creation (learning), and interactions among actors, that is, how knowledge is transferred and built upon (Lundvall, 2007). The strong global links make supernational activities important as well—but global innovation is best viewed as a system of interlinked national systems (Chaminade et al., 2018). This project uses these core concepts from the NIS literature to analyze the evolution of PV.

Energy technology innovation systems

Specific to energy technologies, the "Energy Technology Innovation System" (ETIS) framework emerged from the amalgamation of a wide variety of case studies on energy technology innovation, originating most directly from work by Grubler (1998). It is informed by case study work on innovation more generally (Mowery and Rosenberg, 1998; Arthur, 2009), as well as by energy specific work (Cohen and Noll, 1991; Norberg-Bohm, 2000; Taylor et al., 2003). It also builds on the literature about "innovation systems" (Carlsson and Stankiewicz, 1991; Hekkert et al., 2007; Bergek et al., 2015), which assesses the health of innovation systems by how well a set of functions (e.g. resource mobilization, niche support) operate. As in NIS, interactions among actors are central. A fundamental concept of ETIS is to distinguish it from the linear model of innovation in which science leads to useful innovations (Gallagher et al., 2012). Rather than viewing innovation as a sequence of steps, this perspective emphasizes feedbacks between stages. Both supply and demand for innovation matter; importantly, they interact (Nemet, 2009a). Throughout this book the terms "technology-push" and "demand-pull" are used. The emphasis from ETIS on the need for holistic analysis provides strong motivation for the mixed methods, international, and historical analysis in this project (Grubler and Wilson, 2014).

How it happened, in short

Drawing on the theoretical concepts of "National Innovation Systems" and the "Energy Technology Innovation System," the first question this book asks is: *How did solar get cheap?* In short: inexpensive solar power is the result of a sequence of disparate activities over the past 70 years that involved strong global links, important local activities, and the participation of multiple governments, firms, and influential individuals. It included both technology-push and demand-pull policy instruments. Diverse national systems of innovation complemented each other, sometimes concurrently but mostly sequentially. Progress was not linear, but rather occurred in fits and starts. Like many other technologies before it, solar's importance has been over-estimated in the near term and under-estimated in the longer term. The overall arc of the path to inexpensive solar has involved the following sequence:

1. Scientific discovery, theoretical foundation, and first application, US 1954 (Chapter 3)
2. US federal technology-push (R&D) funding 1973–1981 (Chapter 4);
3. Japanese niche markets in 1980s and rooftop program in 1990s (Chapter 5);
4. Germany's 200-billion-euro demand-pull subsidy from 2000–12 (Chapter 6);
5. Scale up of manufacturing in China from 2001 onwards (Chapter 7); and
6. Local learning in reducing soft costs in the 2010s (Chapter 8).

No one country did it. There was no dominant strategy. Major players—the US, Japan, and Germany—each relinquished their commanding dominance of the industry at some point, meaning that no country applied persistence for the whole lifecycle of PV development (Figure 1.3).

Expectations about future conditions were crucial to the investments that catalyzed PV's improvements. But those expectations were unstable due to the boom and bust cycles of enthusiasm that have characterized PV's evolution. Despite multiple crashes in interest, the emerging PV supply chain never quite died because a progression of niche markets sustained the industry when policy support was lacking. Satellites, offshore oil rigs, telecommunications, consumer electronics, off-grid homes, and green consumers all created demand that was robust to the vagaries of energy policymaking and the volatility of energy markets. PV's ability to function at a variety of scales—less than a Watt in a calculator to a billion times that size in a utility scale plant—made it suitable in a wide variety of applications. Still niche markets alone were insufficient.

Policy entrepreneurs stepped up and took advantage of policy windows—notably in Germany in the late 1990s, as well as China after 2009. Firms hedged policy risk by operating globally and selling to multiple countries. Firm entrepreneurs were central to the creation of a world beating supply chain in China in the early 2000s. Most of the early founders were Australian citizens, many of whom were originally from China. Wind power served as a helpful precursor technology, providing a test-bed for policies when PV remained too expensive for widespread deployment.

Scientific understanding (Einstein, 1905) is at the core of PV technology. Much like a nuclear chain reaction, using photons to dislodge electrons has a technical elegance to it. PV does not involve fuel, friction, heat transfer, or any moving parts at all. PV emerged not from the hunter-gatherer tradition of finding tarry seeps of

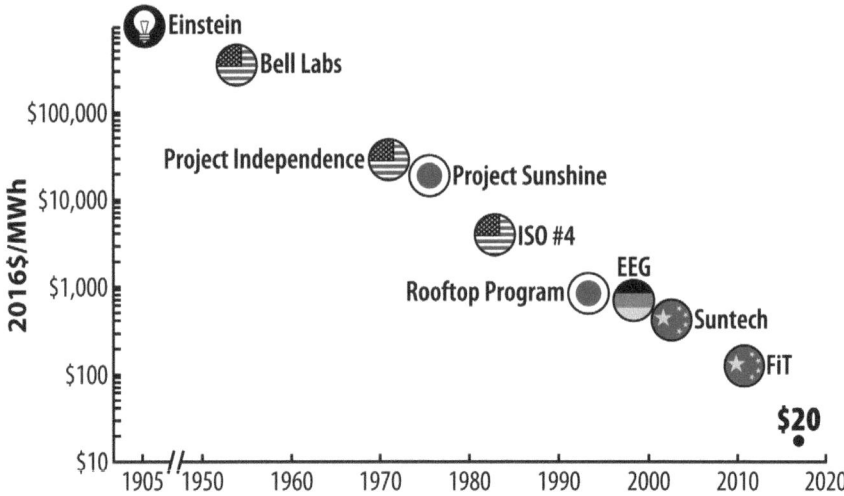

FIGURE 1.3 Milestones in the evolution of inexpensive PV.

oil or burning peaty coal (Yergin, 1991), nor from the experience of feeling the wind and harnessing it (Gipe, 1995). Instead, PV, like nuclear, required the enlightenment. It started with science and has used that understanding of a phenomenon as a basis for continued improvement over decades. Knowledge flowed around the world in the form of scientific publications, conferences, machinery, and above all, people moving from one country to another. PV technology improved through a variety of factors including: early convergence on a dominant design, a clear theoretical foundation in the photoelectric effect, disruptive manufacturing, and implementation of policies oriented toward learning in 1975 (US), in 1994 (Japan), and in 2000 (Germany).

PV as a model

The primary motivation for this book is establishing how PV can serve as a model for other technologies. Understanding how PV got cheap involves looking backwards, documenting the sequence of relevant events, distinguishing important ones from others, and establishing the causal links among them. The real point of doing all that is to search for patterns that may provide a pathway, that is a model, to follow. But a model for what? I contend that PV provides a model for taking an observation of a phenomenon, formalizing it as scientific idea, making it practically useful, and then reducing the costs and scaling it up so that it can compete with and beat gigantic systems that have been optimized over a hundred years. PV has done that against recurring headwinds and long odds.

Knowing what caused the progression in PV helps identify patterns, many of which are pointed to in the innovation literature, that can be applied to others. That PV's sequence has been so overwhelmingly successful is what makes it special and worthy as a case study—not just another case, but a primary one. PV's model will not work for all climate relevant technologies. It will be most useful if we can ex ante know those technologies for which PV serves as an appropriate model and those for which other models are needed. A key question then is: *what are the characteristics of PV that have made it amenable to the successful, albeit too slow, path it has taken?* (Chapter 10). And then: *what aspects of other technologies might make innovation in them function like that of PV?* I find that the following are core characteristics of PV's technological evolution:

- scientific understanding,
- mass production,
- tolerance for design compromises,
- modular scale,
- suitability in niches, and
- opportunities for technology spillovers.

These characteristics help identify other technologies relevant to addressing climate change that could also take this approach. It also helps distinguish those from others that differ in important aspects such that they need other models.

We need a model because a fundamental energy transition requires an array of technologies. Climate change is so big that one technology—even if it is cheap, clean, and reliable—is not enough. Even if it were sufficient, we wouldn't want to rely on just one because there are strong benefits to diversity (Stirling, 1994). At very large scales inherent to the global energy system, very large deployment always creates problems (Grubler, 1998). Such problems can be mitigated by diversification. Work on climate budgets makes clear that we need many decarbonization technologies, applied to many sectors (Rogelj et al., 2018). Arguing about whether we should have 100% renewables or 80% PV misses the point. PV is at about 1% of electricity today. Growing to 50% of electricity or 30% of energy is feasible (Creutzig et al., 2017) and would be massive. But that level also leaves tremendous space for other technologies.

The most important contribution of solar PV's success lies not in gigatons of avoided greenhouse gas emissions, but in establishing a model for successful scale up, from which an array of other climate-related technologies can benefit. The premise in this project is that the success of PV's innovation outcome makes it worth attaining a more complete understanding how it happened. It can increase the chances of successes for the many other technologies we will need to address climate change. As an example of these other technologies, consider negative emissions technologies (NETs) (Minx et al., 2018). Many of these technologies are so new that there is little historical basis from which to form expectations. Using technological analogues has been helpful for informing the technology policy decisions we will soon face on NETs. More completely understanding PV provides an even more useful analogue for some NETs, as well as other technologies, such as small modular nuclear reactors.

Analogous technologies

We can use what we know about PV's success to inform decisions for low-carbon technologies that fit the profile as appropriate analogues to solar (Chapter 10). For example, battery storage is similar to PV in many respects, but is quite far along on its own so doesn't require the guidance of the PV innovation model. I focus on one technology within negative emissions technologies that appears to fit the solar profile quite well: direct air carbon capture and sequestration (DAC). This is an early stage technology with large potential but little experience. Guidance from PV can have a strong impact. Second, I apply the model to small nuclear reactors, which also may be able to learn from aspects of the PV model. Others have potential as well: desalination, fuel cells, high altitude wind, next generation PV, or energy end-use technologies. A key focus here is on how we can accelerate the PV model in applying it to these technologies.

Other technologies important to addressing climate change are not well suited to solar's model—they will need other models. I include this section (within Chapter 10) out of concern that if we only pursue solar-like technologies, we will lose the flexibility that comes from a diverse set of technologies. I see two categories of technologies that are clearly different from PV. First, very large individual units—such as carbon capture at fossil power plants, bioenergy with carbon capture, gigawatt-scale nuclear reactors, and space-based PV—do not provide the opportunities for iterative improvement that solar analogues enjoy. Second, very small low-tech carbon removal methods, such as soil carbon sequestration, afforestation, and enhanced mineralization, do not have the strong potential for production improvements and depend extremely heavily on adoption behavior, continuously over time. Solar analogues are in the middle of these extreme scales. But work shows these technologies have too much potential to dismiss. I offer a perspective on what might be needed to ensure success here with these technologies. That chapter is informed by work on large demonstration projects (Nemet et al., 2018b), as well work on these small NETs (Nemet et al., 2018a).

Accelerating the model

PV's successful evolution is only useful as a model if it can be accelerated. Even though PV's emergence as a serious energy technology seems sudden, it has been a decades-long process. It took 66 years from Becquerel's first observations of photoelectric effects to Einstein's description of the physics underlying the phenomenon. It took another 49 years until Bell Labs used the understanding of photons to design an efficient solar cell. And then a further 63 years until solar was the cheapest form of electricity in the form of $20 per MWh solar contracts and approaching 1% of global electricity production.

That is a key problem to using PV as a model. Even more than the issues with intermittence, financing, and the rest—the slowness of the process despite the impressive accumulation of improvement is the biggest concern. Speed is a problem even though we have ridden down the optimistic path of the learning curve and even though PV has improved more than any other energy technology. Indeed, the more important question may not be "how did solar get cheap?" but rather, *"why did it take so long?"* While PV success is worth emulating, the urgency of addressing climate change means we need to develop scale-up pathways that function faster than PV has. The focus thus is on how to speed up the process of innovation.

Transition debates

The question about how fast energy transitions can occur is a live debate in the academic literature on energy transitions. Some offer a quite optimistic view of adoption, for example by focusing on end-use devices and incremental rather than radical innovation (Sovacool, 2016; Sovacool and Geels, 2016). Others point out that using the historical experience of a broad set of technologies, with consistent measurement of rates of change, suggests that transitions take decades (Grubler et al., 2016). Those focusing on developing countries, whence most new demand for energy will come, point to the risks of locking in to energy-intensive development pathways during the rapid process of industrialization (Fouquet, 2016). An important contribution has been an emphasis on the human dimensions of transitions, since they are central to adoption decisions, use, and the political decisions that affect adoption (Geels et al., 2017). This analysis seeks to make use of the insights in this debate, especially those that are rooted in empirical analysis of other technologies. This theoretical background helps assess whether strategies for accelerating innovation are consistent with this transitions literature. For the PV case, we need to figure out why it took so long.

Expectations

One possible reason that PV developed so slowly is the accumulation of multiple false dawns described above. Perhaps the technology hype cycle that portrayed solar as the answer to energy problems too soon, diminished interest and investment once those hopes were dashed. Solar was hyped to high levels in prominent discussions in the mid-1950s, the late 1970s, and post-2004. The memories of false dawns persisted. For example, Lee Raymond as CEO of Exxon often cited his direct role in winding down the company's solar business in the 1980s as a reason not to take solar seriously (Sobin, 2007). It may have taken even more effort, and thus more expensive subsidies, to overcome the solar skepticism that set in amidst the failed promises of the past. Even today, there is caution against solar following the path of nuclear triumphalism in the 1950s with its "too cheap to meter" claim (Allen and Nemet, 2017; Sivaram, 2018a). A more reasonable set of expectations might have made solar's progress steadier and ultimately faster.

Political economy

Another explanation of solar's slow progress is that it faced opposition from entrenched interests that were politically powerful and effectively slowed its adoption. One sees this in the competition between solar and nuclear power in the 1950s and again in the 1970s (Laird, 2001). Even though the military strongly embraced solar from its earliest days, it became ideologically associated with the left and thus was often discussed on political grounds rather than on the social benefits it might provide (Perlin, 2013). The role of recalcitrant electric utilities that opposed even small amounts of solar is a surprisingly global practice, playing a large role in slowing solar in the US, Japan, and China (Hager and Stefes, 2016). The one place that utilities did not succeed in slowing solar, at least initially, was in Germany, where the bottom-up grassroots support for solar had the upper hand on policymaking (Morris and Jungjohann, 2016). Ultimately, even in Germany, beginning in 2010 utilities and others forced a substantial reform of PV support. One interpretation is that utilities in those other countries learned from the German experience of utility bankruptcies and fought to avoid similar fates by limiting solar adoption through regulatory channels. Solar's progress could have been accelerated if it had acknowledged the influence of utility opposition more seriously and through either confronting them more effectively or, alternatively, finding opportunities for them to benefit from solar.

Leapfrogging

If one country had pursued the entire lifecycle of PV's development, perhaps we could have gotten cheap PV much sooner. I am reluctant to argue that consistent national leadership provides the answer. The national innovation systems literature emphasizes distinct innovation environments. PV drew on features of each of the four leading countries. Knowledge flows were easy and abundant. Individuals moving around the world facilitated global flows of knowledge. One country's model might not have done as well and might even have been slower, for example, if new capabilities had to be developed outside of a country's strengths. Each country's impatience and lack of commitment forced the resulting diversity of incomplete approaches.

That no single country has been the world leader in PV innovation raises the question of whether there exists a self-defeating aspect of being a leader in one facet of technology development. Sure, increasing returns have been important; they tend to reinforce the position of the leader (Arthur, 1989). But they were not deterministic. The torch of PV leadership kept getting passed. Why? Duplication and copying have been important to knowledge diffusion. But also important is adaptation of knowledge to a new context, to a different NIS. Sometimes that adaptation involves aspects that make the innovation perform better than in the original NIS, at least after a while. Perhaps there are "advantages of backwardness" (Morris, 2010) that make catch up and leapfrogging not only possible but inevitable. Technological change in PV meant that what was most important for the next stage of PV's progress shifted over time. Each advantage that an NIS had eventually faded. That is why technology lock-in, even though it was strong, did not dominate the outcomes. To be sure, lock-in played a key role. It produced economies of scale, technological change, and cost reducing improvements. But the extent of that change undermined lock-in's advantages by favoring new combinations of factors and attributes. Globalization enabled PV's markets. But there was more to it than demand. The supply side was also dynamic. Countries could adopt, replicate, and improve based on their unique attributes. Because the technology was changing, the resources needed to progress the technology to the next stage of development also changed. Technological change itself can reshuffle where the advantages are.

My interpretation is that accomplishing all this all in one place would have taken more time not less. I expect something similar will take place for newer climate technologies. We need them to improve. As innovation occurs in a technology, it changes, and it becomes more likely that new, different, and complementary NISs will be needed to sustain innovation. Robust global knowledge flows are needed to take advantage of the diverse contributions that will emerge from existing NISs.

How to speed up the PV model

In Part 4 of this book, I argue that there are at least nine ways to accelerate the PV model so that it can be applied to other technologies in a way that is fast enough for them to contribute meaningfully to climate change. I refer to these as nine *innovation accelerators*:

1. Continuous R&D
2. Public procurement
3. Trained workforce
4. Codify knowledge
5. Disruptive production
6. Robust markets
7. Knowledge spillovers
8. Global mobility
9. Political economy

I discuss these levers in detail and apply them to candidate technologies in Chapter 10.

Organization of this book

How Solar Became Cheap has four parts. Part 1, "*Creating a technology*" covers the origins of PV from the observed phenomenon in the 1800s to the heyday of solar R&D in the US through the early-1980s. I describe how PV evolved from an observation of a strange effect to a Nobel prize winning theory on how it worked to a potential answer to the energy crises of the 1970s. I focus on four key developments: 1) The first solar cell in 1954 and its links to Einstein's 1905 theory of light, 2) the adoption of PV by the US military in the 1950s, 3) the 1970s oil crises, focusing events of which attracted people to work on solar, and 4) government procurement Block Buys, which stimulated early manufacturing.

Part 2, *"Building a market"* emphasizes the transition to a new phase in PV's development. Policy efforts shifted to increase demand for PV electricity. Meanwhile, the private sector continued to find its own new sources of demand, from consumer electronics. Milestones include: the first residential solar subsidy program in Japan in 1993 and the feed-in tariff in Germany in 2000.

Part 3 *"Making it cheap"* covers the period from 2000 to 2017. The role of China is central. But it also describes the local learning that led to reductions in solar electricity production costs other than the hardware which now dominate the total cost. It also discusses the imminent challenges of system integration, especially in developing countries which encounter more difficulty in financing up front investments with multi-year payback periods.

Part 4 *"Doing it again"* returns to the original discussion. How can we learn from PV's learning? First, I clarify the multiple factors that led to inexpensive PV. I then use the outcome to characterize PV and specifically what about PV that made it amenable to these factors. I specify the nine innovation accelerators described above and apply them to do candidate PV analogues. I also discuss climate technologies that differ in important aspects from PV such that they need other models. Next, in Chapter 2, I provide an overview answer to the overall question, *how did solar become cheap?*

References

Allen, T. & Nemet, G. F. 2017. Energy technology exuberance: How a little humility is good for nuclear, renewables, and society. *Medium*.

Anadon, L. D., Bunn, M., Chan, G., Chan, M., Jones, C., Kempener, R., Lee, A., Logar, N. & Narayanamurti, V. 2011. *Transforming U.S. Energy Innovation*. Cambridge, MA: Belfer Center for Science and International Affairs, Harvard Kennedy School.

Anadon, L. D., Baker, E. & Bosetti, V. 2017. Integrating uncertainty into public energy research and development decisions. *Nature Energy*, 2, 17071. Arthur, W. B. 1989. Competing technologies, increasing returns, and lock-in by historical events. *The Economic Journal*, 99, 116–131.

Arthur, W. B. 2009. *The Nature of Technology: What It Is and How It Evolves*, New York, Free Press.

Ausubel, J. H. & Waggoner, P. E. 2008. Dematerialization: Variety, caution, and persistence. *Proceedings of the National Academy of Sciences*, 105, 12774–12779.

Baker, E., Chon, H. & Keisler, J. 2009. Advanced solar R&D: Combining economic analysis with expert elicitations to inform climate policy. *Energy Economics*, 31, S37–S49.

Barbose, G. L., Darghouth, N. R., Millstein, D., Lacommare, K. H., Disanti, N. & Widiss, R. 2017. *Tracking the Sun 10: The Installed Price of Residential and Non-Residential Photovoltaic Systems in the United States*, Lawrence Berkeley National Laboratory and SunShot.

Bergek, A., Jacobsson, S., Carlsson, B., Lindmark, S. & Rickne, A. 2008. Analyzing the functional dynamics of technological innovation systems: A scheme of analysis. *Research Policy*, 37, 407–429.

Bergek, A., Hekkert, M., Jacobsson, S., Markard, J., Sandén, B. & Truffer, B. 2015. Technological innovation systems in contexts: Conceptualizing contextual structures and interaction dynamics . *Environmental Innovation and Societal Transitions*, 16, 51–64. Bleda, M. & Del Rio, P. 2013. The market failure and the systemic failure rationales in technological innovation systems. *Research Policy*, 42, 1039–1052.

Bolinger, M. & Seel, J. 2015. *Utility-Scale Solar 2014: An Empirical Analysis of Project Cost, Performance, and Pricing Trends in the United States*, Lawrence Berkeley National Laboratory.

Bosetti, V., Catenacci, M., Fiorese, G. & Verdolini, E. 2012. The future prospect of PV and CSP solar technologies: An expert elicitation survey. *Energy Policy*, 49, 308–317.

Carlsson, B. & Stankiewicz, R. 1991. On the nature, function and composition of technological systems. *Journal of Evolutionary Economics*, 1, 93–118.

Chaminade, C., Lundvall, B.-Å. & Haneef, S. 2018. *Advanced Introduction to National Innovation Systems*, Cheltenham, Edward Elgar Publishing.

Cohen, L. R. & Noll, R. G. 1991. *The Technology Pork Barrel*, Washington, Brookings.

Craig, P. P., Gadgil, A. & Koomey, J. 2002. What can history teach us: A retrospective examination of long-term energy forecasts for the United States. *Annual Review of Energy and Environment*, 27, 83–118.

Creutzig, F., Agoston, P., Goldschmidt, J. C., Luderer, G., Nemet, G. & Pietzcker, R. C. 2017. The underestimated potential of solar energy to mitigate climate change. *Nature Energy*, 2, nenergy2017140.

Curtright, A. E., Morgan, M. G. & Keith, D. W. 2008. Expert assessments of future photovoltaic technologies. *Environmental Science & Technology*, 42, 9031–9038.

Davidsson, S. & Höök, M. 2017. Material requirements and availability for multi-terawatt deployment of photovoltaics. *Energy Policy*, 108, 574–582.

Davis, S. J. & Fries, K. 2011. *RE: Solar PV Learning Rate: An Expert Discussion*.

Deetjen, T. A., Garrison, J. B., Rhodes, J. D. & Webber, M. E. 2016. Solar PV integration cost variation due to array orientation and geographic location in the Electric Reliability Council of Texas. *Applied Energy*, 180, 607–616.

Dipaola, A. 2017. Saudi Arabia gets cheapest bids for solar power in auction. *Bloomberg Markets* [Online]. Available: https://www.bloomberg.com/news/articles/2017-10-03/saudi-arabia-gets-cheapest-ever-bids-for-solar-power-in-auction [Accessed 4/9/2018].

Dooley, J. J. 1998. Unintended consequences: Energy R&D in a deregulated energy market. *Energy Policy*, 26, 547–555.

Downs, A. 1972. Up and down with ecology: The "issue-attention cycle". *Public Interest*, 28, 38–50.

Einstein, A. 1905. Über einen die Erzeugung und Verwandlung des Lichtes betreffenden heuristischen Gesichtspunkt. *Annalen der Physik*, 322, 132–148.

Farmer, J. D. & Lafond, F. 2016. How predictable is technological progress? *Research Policy*, 45, 647–665.

Fouquet, R. 2016. Lessons from energy history for climate policy: Technological change, demand and economic development. *Energy Research & Social Science*, 22, 79–93.

Freeman, C. 1995. The 'National System of Innovation' in historical perspective. *Cambridge Journal of Economics*, 19, 5–24.

Friedman, B., Ardani, K., Feldman, D., Citron, R., Margolis, R. & Zuboy, J. 2013. *Benchmarking Non-Hardware Balance-of-System (Soft) Costs for US Photovoltaic Systems Using a Bottom-Up Approach and Installer Survey*, NREL.

Fthenakis, V. M. & Moskowitz, P. D. 2000. Photovoltaics: Environmental, health and safety issues and perspectives. *Progress in Photovoltaics*, 8, 27–38.

Gallagher, K. S., Anadon, L. D., Kempener, R. & Wilson, C. 2011. Trends in investments in global energy research, development, and demonstration. *Wiley Interdisciplinary Reviews: Climate Change*, 2, 373–396.

Gallagher, K. S., Grubler, A., Kuhl, L., Nemet, G. & Wilson, C. 2012. The Energy Technology Innovation System. *Annual Review of Environment and Resources*, 37, 137–162.

GEA 2012. *Global Energy Assessment – Toward a Sustainable Future*, Cambridge, UK and New York, Cambridge University Press and Laxenburg, Austria, International Institute for Applied Systems Analysis,.

Geels, F. W., Sovacool, B. K., Schwanen, T. & Sorrell, S. 2017. Sociotechnical transitions for deep decarbonization. *Science*, 357, 1242–1244.

Gillingham, K., Deng, H., Wiser, R. H., Darghouth, N., Nemet, G., Barbose, G. L., Rai, V. & Dong, C. 2016. Deconstructing solar photovoltaic pricing: The role of market structure, technology, and policy. *The Energy Journal*, 37, 231–250.

Gipe, P. 1995. *Wind Energy Comes of Age*, New York: Wiley.

Gowrisankaran, G., Reynolds, S. S. & Samano, M. 2016. Intermittency and the value of renewable energy. *Journal of Political Economy*, 124, 1187–1234.

Graziano, M. & Gillingham, K. 2015. Spatial patterns of solar photovoltaic system adoption: The influence of neighbors and the built environment. *Journal of Economic Geography*, 15, 815–839.

Green, M. A., Emery, K., Hishikawa, Y., Warta, W. & Dunlop, E. D. 2016. Solar cell efficiency tables (version 48). *Progress in Photovoltaics: Research and Applications*, 24, 905–913.

GREENPEACE 2015. *A Sustainable World Energy Outlook 2015*.

Grubler, A. 1998. *Technology and Global Change*, Cambridge, Cambridge University Press.

Grubler, A. & Wilson, C. 2014. *Energy Technology Innovation: Learning from Historical Successes and Failures*, Cambridge, Cambridge University Press.

Grubler, A., Wilson, C. & Nemet, G. 2016. Apples, oranges, and consistent comparisons of the temporal dynamics of energy transitions. *Energy Research and Social Science*, 22, 18–25.

Hager, C. & Stefes, C. H. 2016. *Germany's Energy Transition: A Comparative Perspective*, Springer.

Hekkert, M. P., Suurs, R. A. A., Negro, S. O., Kuhlmann, S. & Smits, R. 2007. Functions of innovation systems: A new approach for analysing technological change. *Technological Forecasting and Social Change*, 74, 413–432.

Hertwich, E. G., Gibon, T., Bouman, E. A., Arvesen, A., Suh, S., Heath, G. A., Bergesen, J. D., Ramirez, A., Vega, M. I. & Shi, L. 2015. Integrated life-cycle assessment of electricity-supply scenarios confirms global environmental benefit of low-carbon technologies. *Proceedings of the National Academy of Sciences*, 112, 6277–6282.

Herzog, H. J. 2011. Scaling up carbon dioxide capture and storage: From megatons to gigatons. *Energy Economics*, 33, 597–604.

Holdren, J. P. & Baldwin, S. F. 2001. The PCAST energy studies: Toward a national consensus on energy research, development, demonstration, and deployment policy. *Annual Review of Energy and Environment*, 26, 391–434.

Hoppmann, J., Huenteler, J. & Girod, B. 2014. Compulsive policy-making—The evolution of the German feed-in tariff system for solar photovoltaic power. *Research Policy*, 43, 1422–1441.

Hutchby, J. A. 2014. A Moore's Law-like approach to roadmapping photovoltaic technologies. *Renewable and Sustainable Energy Reviews*, 29, 883–890.

IPCC 2011. *The IPCC Special Report on Renewable Energy Sources and Climate Change Mitigation*. Intergovernmental Panel on Climate Change (IPCC).

IRENA 2017. *Renewable energy auctions: Analyzing 2017.*
IRENA 2018. *Renewable Power Generation Costs in 2017.*
Jackson, R. B., Quéré, C. L., Andrew, R. M., Canadell, J. G., Peters, G. P., Roy, J. & Wu, L. 2017. Warning signs for stabilizing global CO_2 emissions. *Environmental Research Letters*, 12, 110202.
Klein, N. 2015. *This Changes Everything: Capitalism vs. the Climate*, New York, Simon & Schuster.
Knapp, K. E. 1999. Exploring energy technology substitution for reducing atmospheric carbon emissions. *The Energy Journal*, 20, 121–143.
Laird, F. N. 2001. *Solar Energy, Technology Policy, and Institutional Values*, New York, Cambridge University Press.
LAZARD 2018. *Lazard's Levelized Cost of Energy Analysis, Version 12.0.*
List, F. 1841. *Das nationale System der politischen Oekonomie.* Waentig Heinrich, Sammlung sc.
Lundvall, B.-A. (ed.) 1992. *National Innovation System: Towards a Theory of Innovation and Interactive Learning*, London, Pinter.
Lundvall, B. Å. 2007. National innovation systems—analytical concept and development tool. *Industry and Innovation*, 14, 95–119.
Margolis, R. M. & Kammen, D. M. 1999. Underinvestment: The energy R&D challenge. *Science*, 285, 690–692.
McDonald, A. & Schrattenholzer, L. 2001. Learning rates for energy technologies. *Energy Policy*, 29, 255–261.
Millar, R. J., Fuglestvedt, J. S., Friedlingstein, P., Rogelj, J., Grubb, M. J., Matthews, H. D., Skeie, R. B., Forster, P. M., Frame, D. J. & Allen, M. R. 2017. Emission budgets and pathways consistent with limiting warming to 1.5 °C. *Nature Geoscience*, 10, 741–747.
Minx, J. C., Lamb, W. F., Callaghan, M. W., Fuss, S., Hilaire, J., Creutzig, F., Amann, T., Beringer, T., Garcia, W. D. O., Hartmann, J., Khanna, T., Lenzi, D., Luderer, G., Nemet, G. F., Rogelj, J., Smith, P., Vicente, J. L. V., Wilcox, J. & Dominguez, M. D. M. Z. 2018. Negative emissions: Part 1—research landscape and synthesis. *Environmental Research Letters*, 13, 063001.
Morris, C. & Jungjohann, A. 2016. *Energy Democracy: Germany's Energiewende to Renewables*, Springer.
Morris, I. 2010. *Why the West Rules—For Now: The Patterns of History and What They Reveal About the Future*, London, Profile Books.
Mowery, D. C. & Rosenberg, N. 1998. *Paths of Innovation: Technological Change in 20th-Century America*, Cambridge, Cambridge University Press.
Nelson, R. 1993. *National Innovation Systems. A Comparative Analysis*, Oxford, Oxford University Press.
Nemet, G. F. 2009a. Demand-pull, technology-push, and government-led incentives for non-incremental technical change. *Research Policy*, 38, 700–709.
Nemet, G. F. 2009b. Net radiative forcing from widespread deployment of photovoltaics. *Environmental Science & Technology*, 43, 2173–2178.
Nemet, G. F. 2013. Technological change and climate-change policy. In: Shogren, J. (ed.) *Encyclopedia of Energy, Natural Resource and Environmental Economics*. Amsterdam: Elsevier.
Nemet, G. F. & Kammen, D. M. 2007. U.S. energy research and development: Declining investment, increasing need, and the feasibility of expansion. *Energy Policy*, 35, 746–755.
Nemet, G. F., Braden, P., Cubero, E. & Rimal, B. 2014. Four decades of multiyear targets in energy policy: Aspirations or credible commitments? *Wiley Interdisciplinary Reviews: Energy and Environment*, 3, 522–533.
Nemet, G. F., Grubler, A. & Kammen, D. 2016. Countercyclical energy and climate policy for the U.S. *Wiley Interdisciplinary Reviews: Climate Change*, 7, 5–12.

Nemet, G. F., Callaghan, M. W., Creuzig, F., Fuss, S., Hartmann, J., Hilaire, J., Lamb, W. F., Minx, J. C., Rogers, S. & Smith, P. 2018a. Negative emissions – Part 3: Innovation and upscaling. *Environmental Research Reviews*, 13, 063003.

Nemet, G. F., Zipperer, V. & Kraus, M. 2018b. The valley of death, the technology pork barrel, and public support for large demonstration projects. *Energy Policy*, 119, 154–167.

Noailly, J. & Shestalova, V. 2017. Knowledge spillovers from renewable energy technologies: Lessons from patent citations. *Environmental Innovation and Societal Transitions*, 22, 1–14.

Norberg-Bohm, V. 2000. Creating incentives for environmentally enhancing technological change: Lessons from 30 years of US energy technology policy. *Technological Forecasting and Social Change*, 65, 125–148.

NRC 2007. *Prospective Evaluation of Applied Energy Research and Development at DOE (Phase Two)*, Washington, The National Academies Press.

O'Shaughnessy, E. 2018. *Solar Photovoltaic Market Structure in the United States: The Installation Industry, Effects on Prices, and the Role of Public Policy*. PhD, University of Wisconsin-Madison.

PCAST 1997. *Report to the President on Federal Energy Research and Development for the Challenges of the Twenty-First Century*, Washington, Office of the President.

PCAST 1999. *Powerful Partnerships: The Federal Role in International Energy Cooperation on Energy Innovation*. Washington: Office of the President.

Perlin, J. 1999. *From Space to Earth: The Story of Solar Electricity*, Ann Arbor, MI, AATEC Publications.

Perlin, J. 2013. *Let it Shine: The 6,000-year Story of Solar Energy*, Novato, CA, New World Library.

Peters, G. P., Andrew, R. M., Canadell, J. G., Fuss, S., Jackson, R. B., Korsbakken, J. I., Le Quere, C. & Nakicenovic, N. 2017. Key indicators to track current progress and future ambition of the Paris Agreement. *Nature Climate Change*, 7, 118–122.

Pillai, U. 2015. Drivers of cost reduction in solar photovoltaics. *Energy Economics*, 50, 286–293.

Rai, V. & Robinson, S. A. 2013. Effective information channels for reducing costs of environmentally-friendly technologies: Evidence from residential PV markets. *Environmental Research Letters*, 8, 014044.

Rai, V. & Sigrin, B. 2013. Diffusion of environmentally-friendly energy technologies: Buy versus lease differences in residential PV markets. *Environmental Research Letters*, 8, 014022.

Rogelj, J. 2017. Transition pathways towards limiting climate change below 1.5°C. under review at *Nature Climate Change*.

Rogelj, J., Popp, A., Calvin, K., Luderer, G., Emmerling, J., Gernaat, D. et al., 2018. Scenarios towards limiting climate change below 1.5°C. *Nature Climate Change*, 8, 325–332.

Pielke, R., Jr. 2010. *The Climate Fix: What Scientists and Politicians Won't Tell You About Global Warming*, New York, Basic Books.

Rosenberg, N. 1990. Why do firms do basic research (with their own money)? *Research Policy*, 19, 165–174.

Rubin, E. S., Azevedo, I. M. L., Jaramillo, P. & Yeh, S. 2015. A review of learning rates for electricity supply technologies. *Energy Policy*, 86, 198–218.

Ryan, S. E., Hebdon, C. & Dafoe, J. 2014. Energy research and the contributions of the social sciences: A contemporary examination. *Energy Research & Social Science*, 3, 186–197.

Schock, R. N., Fulkerson, W., Brown, M. L., Martin, R. L. S., Greene, D. L. & Edmonds, J. 1999. How much is energy research and development worth as insurance? *Annual Review of Energy and Environment*, 24, 487–512.

Sivaram, V. 2018a. A tale of two technologies: What nuclear's past might tell us about solar's future. *The Breakthrough Journal*, 8.

Sivaram, V. 2018b. *Taming the Sun: Innovations to Harness Solar Energy and Power the Planet*, Cambridge, MA, The MIT Press.

Smil, V. 2008. Long-range energy forecasts are no more than fairy tales. *Nature*, 453, 154–154.
Sobin, R. 2007. Energy myth seven—Renewable energy systems could never meet growing electricity demand in America. In: Sovacool, B. K. & Brown, M. A. (eds.) *Energy and American Society – Thirteen Myths*. Springer.
Sovacool, B. K. 2016. How long will it take? Conceptualizing the temporal dynamics of energy transitions. *Energy Research & Social Science*, 13, 202–215.
Sovacool, B. K. & Geels, F. W. 2016. Further reflections on the temporality of energy transitions: A response to critics. *Energy Research \& Social Science*, 22, 232–237.
Stirling, A. 1994. Diversity and ignorance in electricity supply investment. *Energy Policy*, 22, 195–216.
Taylor, M. R., Rubin, E. S. & Hounshell, D. A. 2003. Effect of government actions on technological innovation for SO2 control. *Environmental Science & Technology*, 37, 4527–4534.
Teece, D. J. 1986. Profiting from technological innovation – implications for integration, collaboration, licensing and public-policy. *Research Policy*, 15, 285–305.
Tybout, R. A. 1972. Pricing pollution and other negative externalities. *Bell Journal of Economics and Management Science*, 3, 252–266.
UNFCCC 2015. *The Paris Agreement. United Framework Convention on Climate Change*.
Verdolini, E., Anadon, L. D., Lu, J. & Nemet, G. F. 2015. The effects of expert selection, elicitation design, and R&D assumptions on experts' estimates of the future costs of photovoltaics. *Energy Policy*, 80, 233–243.
Verdolini, E., Anadón, L. D., Baker, E., Bosetti, V. & Aleluia Reis, L. 2018. Future prospects for energy technologies: Insights from expert elicitations. *Review of Environmental Economics and Policy*, 12, 133–153.
Wiser, R. H. & Bolinger, M. 2017. 2016 *Wind Technologies Market Report*.
Wolfe, P. R. 2018. *The Solar Generation: Childhood And Adolescence Of Terrestrial Photovoltaics*, Hoboken, NJ, IEEE Press & Wiley.
Yergin, D. 1991. *The Prize: The Epic Quest for Oil, Money, and Power*, New York, Simon and Schuster.
Zhang, F. & Gallagher, K. S. 2016. Innovation and technology transfer through global value chains: Evidence from China's PV industry. *Energy Policy*, 94, 191–203.

2
ANSWER

Solar became inexpensive through a progression of contributions: first from the US, then Japan, then Germany, Australia, and finally China. The torch of PV leadership passed from one country to another. Expectations about future technological opportunities and future market opportunities were crucial to investment. But they were also were unstable due to policy volatility, as well as recurring disappointment about near term technological progress and adoption. Niche markets, sets of consumers who had a willingness to pay for PV applications at prices above those of grid-electricity, kept the PV industry alive when policy support was lacking. The modular aspect of PV—it could function at a wide variety of scales—provided a richer set of niche markets than could a technology with a narrower range of unit sizes. PV's human scale also contributed to a public enthusiasm towards solar which helped sustain policy support, a relationship not commonly seen with other energy technologies.

In short, PV improved as the result of:

1. Scientific contributions in the 1800s and early 1900s, in Europe and the US, that provided a fundamental understanding of the ways that light interacts with molecular structures (Chapter 3);
2. A breakthrough at a corporate laboratory in the US in 1954 that made a commercially available PV device (Chapter 3);
3. A major government R&D and public procurement effort in the 1970s in the US (Chapter 4);
4. Japanese electronic conglomerates serving niche markets in the 1980s and in 1994 launching the world's first major rooftop subsidy program, with a declining rebate schedule (Chapter 5);
5. Germany passing a feed-in tariff in 2000 that quadrupled the market for PV and developing production equipment that automated and scaled PV manufacturing (Chapter 6);

6. Chinese entrepreneurs, trained in Australia, building factories of gigawatt scale in the 2000s and creating the world's largest market for PV from 2013 onward (Chapter 7); and
7. A cohort of adopters with high willingness to pay, accessing information from neighbors, and installer firms that learned from their installation experience, as well as that of their competitors to lower soft costs (Chapter 8).

Policymakers stepped up with support for PV at crucial times. Often these efforts depended on a small number of policy entrepreneurs who had championed PV for years and took advantage of the emergence of a policy window. Other times promising policy windows were missed. Wind power provided a helpful precursor technology in that policies and systems could be tested with wind power while PV was still too expensive to commercially deploy. For example, government price guarantees in California in the mid-1980s and Germany in the early 2000s were applied to wind before PV. By the time PV was ready for support, policy design had matured and confidence in its mechanisms had increased.

Knowledge flowed easily across international borders; publications, conferences, machines, and individuals moved globally. These global knowledge flows enabled new contributors to emerge and make use of their distinct national innovation systems. PV technology benefited from a theoretical foundation established by Albert Einstein in 1905, early convergence on a dominant design, and spillovers from the computer industry. Later, these factors supported China's extensive entrepreneurial efforts to massively scale up PV manufacturing, which in turn made it inexpensive and accessible. In the early 2000s experience in semi-conductor manufacturing was crucial for PV development, but as costs fell, China was able to take full advantage of its experience in manufacturing lower technology commodity products, an industry where profit margins are small and scale was the key to profit. After 2010, PV became a brutally competitive low margin commodity business.

The challenges of integrating intermittent PV into an electric grid designed for dispatchable fossil fuel generation still remain. But PV is in a much better position to address those issues now that it is inexpensive and has a substantial industrial base behind it. Had PV needed to deal with integration of a complex system when it was nascent and expensive, it would have been much less likely to have catalyzed the positive feedback between deployment and learning. By 2018, 13 countries were installing over a GW per year of new PV each. PV had become among the least expensive ways to generate electricity.

Passing the torch

The first explanation in tracking the progress in solar is that no single country led its development. Leadership in PV was like a relay race, with each leader passing the torch to another, sometimes involuntarily. One way to measure leadership is by looking at global market share in PV manufacturing (Figure 2.1). The US led in the 1970s when it accounted for most of world production, but then Japan took

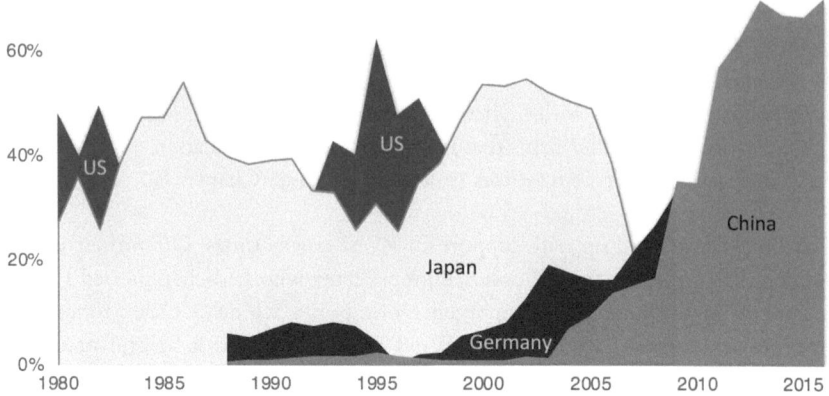

FIGURE 2.1 Share of world PV production.

over market control by the mid-1980s. A decade later, US had regained the role of top producer, just to be replaced again by Japan's lead and eventual peak in 2004. Industry leadership shifted for two years to Germany before China overtook it from then on. By 2011 China's share surpassed any previous country's share, with the possible exception of US dominance in the 1970s when comprehensive data are less reliable.

Looking at the national share of demand—in the form of installations of PV rather than production—produces a slightly different perspective. We still see the eras of US, Japanese, and Chinese dominance, but now also see Germany's important role in the 2004–2011 period (Table 2.1). Germany's role as a producer of PV systems, despite great ambitions in the mid-2000s, never matched its role as a consumer of solar panels, and thereby as a generator of solar electricity. To be sure, Germany led the world in making PV production equipment, selling tens of billions of dollars' worth of equipment to Chinese companies between 2005 and 2011. In contrast, China did not pursue a large domestic market until it had established itself as a dominant producer (Figure 2.2).

An important implication of these cycles of national competitiveness is that knowledge about PV had to move from one country to another. Progress in PV depended on each country not starting anew but building on the efforts of precursor countries. This is where the national system of innovation perspective is most helpful to understanding solar's evolution. Each time a transition occurred, the new leader was able to absorb the extant knowledge, and then add its own distinct contribution.

Overlapping leadership

The national innovation systems (NIS) perspective is useful for understanding why PV proceeded in this sequence of overlapping phases of leadership. In short, each country's national system of innovation allowed it to make a unique contribution to the evolution of solar. No country continuously persisted in pushing solar

TABLE 2.1 Shares of world PV manufacturing and installations.

	Manufacturing			Installations		
	Year of 1st lead	Peak year	Peak share	Year of 1st lead	Peak year	Peak share
U.S	1957	1995	63%	1957	1957	100%
Japan	1985	2002	55%	1995	1998	73%
Germany	2007	2010	35%	2004	2005	67%
China	2009	2016	71%	2013	2016	48%

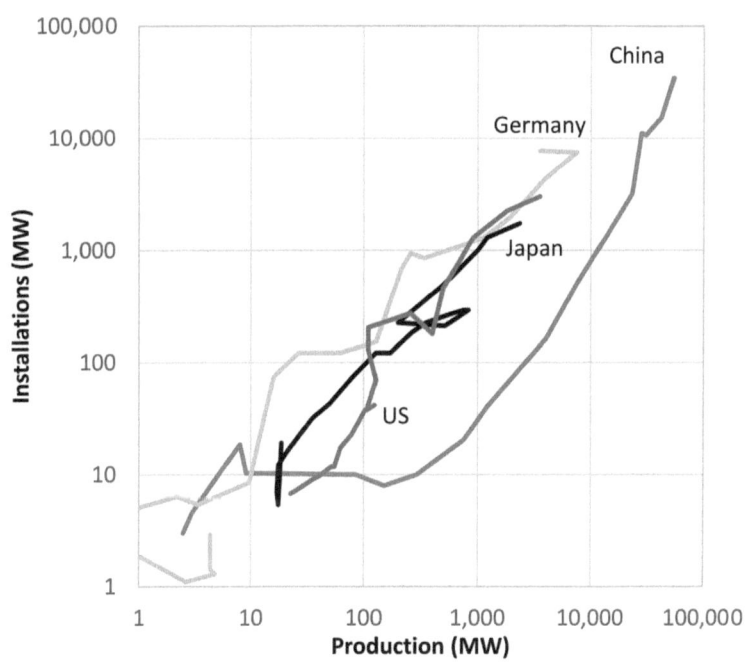

FIGURE 2.2 Installations and production by the four leading PV actors 1992–2016.

forward. No single country had the capability or the perseverance to take solar from its origin to widespread adoption. Repeatedly, when favorable policies ended, those in other jurisdictions stepped up—sometimes governments at the subnational level and other times national governments of other countries. Each country was able to build on the work of the earlier leaders. Knowledge flows were easy and abundant as individuals that moved around the world enabled knowledge to move globally rapidly.

The NIS perspective would argue that understanding how solar got cheap requires a focus on knowledge and learning. It says that creating and acquiring new knowledge works differently in different national contexts. There are different modes of innovation, which complement each other. NIS includes important roles both for the actors and the surrounding context. For example, in the solar industry, actors would include scientists, households adopting solar, entrepreneurial firms, and the people who work in them. The surrounding context involves institutions, policymakers, and the ways in which the state interacts with the actors above. Educational systems and culture are also important elements of the NIS perspective. The NIS argues that an innovation system is effective when these various actors "fit" and are "aligned" with each other. Throughout the literature there is an emphasis on understanding the extent to which actors and the surrounding context support the fundamental objective within the NIS perspective, learning.

> **BOX 2.1 INNOVATION TERMS.**
>
> **Innovation:** The process of creating and adopting novel technologies, processes, and behaviors.
> **NIS:** National Innovation Systems is a theory in the literature on innovation that emphasizes the distinct characteristic of the national context in which innovation occurs. Education, institutions, and culture influence innovative activities.
> **ETIS:** The Energy Technology Innovation System emphasizes the feedbacks of knowledge that occur throughout the innovation system.
> **Demand-pull:** Public policies that increase the payoffs to successful innovations, for example by expanding the markets for them.
> **Technology-push:** Public policies that reduce the costs of producing innovations, for example by funding research and development.
> **Niche markets:** Sets of adopters of an innovation that are willing to pay more than those in the larger market to which the innovation is ultimately meant to serve.
> **Scale up:** In the PV context it typically refers to increasing the size of manufacturing facilities. It is also sometimes used to talk about the path to widespread adoption. It can also refer to increasing unit size, for example progressing from 100kW wind turbines to 1MW turbines.
> **Economies of scale:** Reductions in the costs to produce each unit of a good, for example by spreading fixed costs across a larger number of units.

Expectations were crucial

Expectations are central to innovation in solar. So many of the activities—investment, policy adoption, technology development, and career changes—had to be forward-looking because the payoffs were distant. Companies pivoted to solar to serve new markets they identified. Individuals chose or changed their career paths. Policymakers gambled on polices with high near-term costs and uncertain long-term benefits. Investors funded the construction of PV "fabs," fabrication facilities, which involved hundreds of millions of dollars in investment. Supporters of climate change mitigation advocated for policies with time horizons for payoffs that involved many decades. All of these activities were important to improvements in solar, and they all hinged on expectations of the future conditions of the industry.

False dawns and policy windows

As important as expectations were to the decisions that enabled inexpensive solar, it is striking how often those expectations were dashed by disappointment. Bell Lab's announcement of the first solar cell in April 1954 prompted a front page article in the *New York Times*: "New Battery Taps Sun's Vast Power" (NYTimes, 1954). The Eisenhower administration, at one point, considered PV a rival to nuclear (President's Materials Policy Commission, 1952). The second oil crisis of the 1970s, in 1979, led to the "Domestic Policy Review of Solar Energy" and President Carter's plan to meet 20% of energy supply with solar by the year 2000 (United States Department of Energy, 1979). The Chernobyl nuclear accident in 1986 catalyzed efforts in Germany to switch to renewables. Japan launched its New Sunshine program in December 1994 (Tatsuta, 1996) with a plan for 5 GW by 2010. China made PV part of its program of rural electrification in 1995 (Dunford et al., 2013). While eventually each of these plans was achieved or nearly so, all were disappointments in the near term.

BOX 2.2 MILESTONES IN THE HISTORY OF PV.

1839: Becquerel discovers the photoelectric effect.
1905: Einstein publishes Nobel Prize winning theory explaining the photoelectric effect.
1954: Bell Labs announce the first practical solar cell.
1973: President Nixon launches Project Independence. Japan launches Project Sunshine.
1975: US Department of Energy Block Buy Program.
1983: California Interim Standard Offer Contract #4
1985: Oil prices collapse.
1994: Japan launches its Rooftop Subsidy Program.
2001: Suntech is founded.
2004: Germany updates its Renewable Energy Law to be much more favorable to PV.
2005: Suntech IPO and scale up of Chinese firms.
2011: Chinese feed-in tariff.
2018: Multiple PV projects below $20 per MWh. PV is cheap.

Energy is prone to booms and busts (Nemet et al., 2016). Like other parts of the energy sector—in particular oil, gas, and wind—solar's evolution has not been a consistent rise. It has involved an array of setbacks. While solar has had moments of recognizing its immense potential, we have also seen the opposite in every major solar country.

In the US, disappointment in solar occurred multiple times. Truman chose nuclear over solar. Carter gave up on solar in the second half of his presidency. Reagan tried to close the Department of Energy in 1981. Japanese companies got out of solar in the late 1980s. Germany lost much of its manufacturing business in the 2008–11 period and cut its subsidy program radically in 2012. China's early use of solar for rural electrification disappointed. Its solar industry has been awash in red ink and bankruptcies since 2012.

Booms and busts characterize the long-term history of PV. And while several countries have played key roles in the development of solar, a recurring theme in this story is the volatility of each country's interest in solar. R&D funding is the most direct representation of that changing prioritization. One can see this volatility clearly in PV R&D funding in the US (Nemet and Kammen, 2007). This R&D volatility is not unique to PV, it has affected energy overall (Schuelke-Leech, 2014). Recurring waves of investment and disinvestment have passed. For many years Japan was held as a model, espousing the opposite of US R&D volatility. Indeed, its funding was quite stable from 1980 to 2004. But over the complete time horizon even Japan was challenged by unstable funding.

Up until the post-2010 period, PV was always a bit too far away from commercial feasibility to take off. As a result, it is somewhat surprising that PV has not gone extinct, like other technologies that don't find a market. Policymakers, firms, scientists, and the public kept moving on from solar. But somehow solar kept coming back. That is one of its key reasons for success and it was niche markets that kept the PV industry alive.

Niche markets

In part because of the massive scale of the decarbonization challenge discussed above, serving small idiosyncratic markets is often dismissed as proof that a technology is not serious. Yet, niche markets are a central feature in the literature on innovation (Kemp et al., 1998; Raven et al., 2016). Niche markets exist when early adopters have a higher than average willingness to pay for a technology. Such markets have been crucial for solar. Despite widespread characterization of solar as the domain of hippies, scientists, and environmentalists (Laird, 2003), it is striking how some of the key actors in solar's development held none of these attributes. Rather, entrepreneurs like Les Hoffman in the 1950s and large companies like Sharp in the 1980s, focused on taking advantage of a business opportunity. The US military was its main supporter in the 1950s. At various times in its evolution, solar provided a unique way to serve various needs.

Space programs in multiple countries provided important early niche markets for PV. After 60 years of launches, there are a few thousand satellites orbiting the earth. That creates a limited market for PV. But the cost of electricity is expensive and the alternatives to PV, batteries and fuel cells, are expensive. Since sponsors of space programs have much higher willingness to pay for electricity than do residential electricity consumers, they were a perfect candidate for PV adoption. Even if these niche markets are small, they can enable subsequent scale up, which provides opportunities to decrease costs. Niche markets can be important to launching risky new technologies (Kivimaa and Kern, 2015), especially those whose initial costs are high but may fall subsequently through learning by doing. They can also provide temporary insulation from competition, for example when the scales of existing competitors might make them uncompetitive (Kemp et al., 1998; Raven et al., 2016). Some have argued that serving niches diverts innovation toward characteristics that may be useful for the niche but not for large-scale electricity delivery. There is something to that in that consumer electronics stimulated efforts on thin film technologies in the 1980s, whereas about 90% of PV today does not use thin-films. In this case there seems to have been enough spillover among the technologies for the niches to support the industry overall.

A demand curve of niches

The flexibility of solar to function in a wide array of conditions, many that had little to do with the market for bulk electricity, benefited the development of the industry and of the technology. The list of applications for which solar supplied power is lengthy: navigation aids on offshore oil rigs, buoys, lighthouses, telecom repeater stations, satellites, calculators, radios, toys, and off grid houses. Each of these markets was trivial compared to the 1% of electricity supply that solar serves today, never mind the 10–50% that solar could potentially contribute.

However, these niche markets were important because they created a path. They occurred in succession and generally increased in size and decreased in willingness to pay. One could portray these markets as a demand curve. Figure 2.3 is illustrative to convey the idea of a succession of markets. Note the dual logarithmic scales. I originally made this figure in 2005 while in graduate school (Nemet, 2006). I have plotted the price of the 2017 power purchase agreements (PPAs) described in Chapter 1 and one can see it lies below the curve. Like others (Verdolini et al., 2015), I was pessimistic in 2005 about how inexpensive PV could be expected to get. I may also have been pessimistic in not extending the curve further to the right. At the time speaking of annual terawatts seemed fantastical, yet today analysts are seriously considering terawatts of PV installations per year (Haegel et al., 2017; Davidsson and Höök, 2017). If the growth rate of the past 5 years (28%) is sustained, 1 TW of new PV per year will be installed in 2027.

At any size

Solar was able to serve these niche market opportunities in large part because of its flexible scale: it could be deployed at large scale (100s of MWs), small scale (a wristwatch cell is less than a Watt), and anything in between. Utility scale solar is a billion times bigger than a wristwatch cell. Solar is unique in that. So even though it got cheap in large part through economies of scale in manufacturing, and ultimately in economies of scale in large installations, it could be deployed in almost

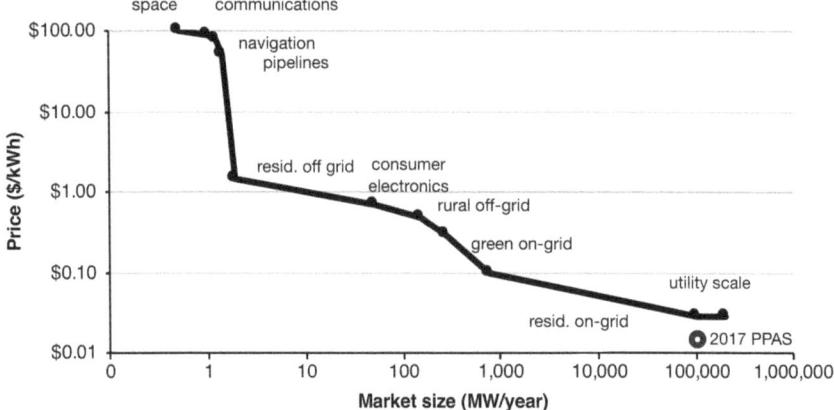

FIGURE 2.3 Illustrative curve of a progression of markets for PV electricity. Circle indicates power purchase agreements (PPAs) signed in 2017.

any size. This modularity enabled not only suitability in a variety of applications but also distributed decision making. Those other than large companies and regulated utilities could participate in and benefit from PV. The flexible scale of PV meant that individuals and entrepreneurs played important roles.

Independent of policy

An additional attribute of the niche-serving path was that these niche opportunities were almost entirely independent of public policies directed at solar itself. To be sure, governments played a role in running the space program, in subsidizing offshore oil exploration, and in supporting the consumer electronics industry. But these government activities were not designed with solar in mind. This independence from policy was a further benefit of niche markets because policy volatility has been such a central aspect of energy policymaking (Nemet et al., 2014). Niches made solar less dependent on policy. Policymakers were essential to solar's eventual success in becoming economically competitive, but a key aspect of its success was solar's ability to survive and grow during the interludes when public interest in solar was low. For example, the existence of niche applications in consumer electronics allowed the PV industry to survive even when policy support was weak after energy prices fell in 1985 and stayed low for two decades.

Policymakers stepped up

Still, public policy has been important to solar's development. To be sure, no single country had the capability or the perseverance to take solar from birth to today's level of widespread adoption; no country persisted. Further, national innovation systems allowed each country to make a unique contribution. Of course, the flip side of the claim that polices were volatile and lacked persistence is that from time to time policymakers stepped up. For example, at times they developed innovative policies, by assembling coalitions to pass legislation, and crucially, by being prepared once a policy window opened.

Policy windows

Political scientists use the notion of "policy windows" and "three streams" in the policymaking process (Kingdon, 1984). When the problem stream (overcoming limited attention), the policy stream (availability of ideas and solutions), and the politics stream (elections and national mood) align favorably, a "window" can open that makes it possible to adopt legislation, which previously had not been politically feasible. In the evolution of solar, there are numerous examples in which solar advocates took advantage of policy windows. Even beyond the direct policy realm, windows of opportunity emerged. These windows were often short, sometimes missed, and occasionally used successfully. Examples of these windows include the launch of Sputnik in 1957, the 1973 Arab Oil Embargo, the 1979 Iranian Revolution, Chernobyl in 1986, Germany's Red-Green coalition in 1998, September 11, 2001 in the US, and the global financial crisis in 2009–10.

Solar benefited from these windows when a framework had been built previously so that the opportunity could be seized quickly. In the strongest version of this interpretation: even if the window was short, acting within it could catalyze enough activity that increasing returns took over. When the window closed, forces supporting the action persisted, making reversal of course difficult. One could think of legislation enshrined in law, economies of scale in production, and the development of an advocacy coalition (Jenkins-Smith et al., 2014) as forces that persisted once the three streams of problem, policy, and politics unraveled and each wandered off on their own. China's response to the 2004 EEG provides a good example. China had, and still has, a labor cost advantage over Germany. Today, with so much automation in solar production, labor cost differences do not provide much advantage (Goodrich et al., 2013). But in 2004, PV manufacturing in China was far less automated. This advantage helped fuel China's scale up in response to the rapid expansion of the German market. Another window for China came in 2008–09 with the global financial crisis. While the credit constraints and uncertain markets strangled expansion elsewhere, China persisted.

In other cases, solar made less progress than anticipated, despite the appearance of windows of opportunity. The hopes for solar after Sputnik and the two oil crises in the 1970s were largely met with disappointment; policymakers, firms, or the technologies themselves were not sufficiently prepared to take advantage of the opportunity. Solar was not part of the technology response to September 11, rather it was hydrogen, "freedom fuel" as President Bush called it in the January 2003 State of the Union address. The instances in which solar took advantage of windows of opportunity are central to its success. The many cases of solar missing windows help contribute to our understanding of why it took so long.

Countries stepped up

From a global perspective, the instances of taking advantage of, rather than missing, windows seem more apparent. Several times, when solar-related policies ended or a country shifted its attention away from solar, other jurisdictions stepped up. In the most global view, as discussed above, we see countries relinquishing and taking up leadership on solar. Indeed, the leadership progression of the US, to Japan, to Germany, to China forms the narrative arc of this book. But it is important to note that other countries played central roles at certain times. Australia led PV R&D efforts in the 1980s as the US program downshifted under Carter and Reagan. Spain was the largest market in the world in 2008 as Germany's growth slowed. Malaysia became a locus of PV production post-2010. The torch passing metaphor involved more than the big four players. Sub-national jurisdictions also played a role in driving progress, even when their own national governments appeared not to play. For example, Baden-Württemberg in southwestern Germany developed a whole cluster of PV activities in the 1980s and 1990s before the federal EEG was passed in 2000. California stepped up in the 1980s as federal budgets were being cut, and again in 2006 when solar was not a favored technology at the US federal level.

Inherently local

A predominant distinguishing characteristic of solar is that even though it depends on a deeply globalized supply chain, a substantial portion of value added has remained local. Even after decades of optimizing solar production, the local content remains large. In fact, as a proportion, the local added value is higher than ever: about two-thirds for rooftop systems (Barbose et al., 2017), and half for utility scale (Bolinger et al., 2017). Using industry jargon, these "balance of system" or "soft costs," involve labor in installation, insurance, marketing, and obtaining permits (Friedman et al., 2013). These costs are distinct from hardware costs in that they are almost completely locally driven (Karakaya et al., 2016), whereas PV modules are globally traded. Using recent data, 80% of the costs for residential systems are soft and thus local, while only 20% are based on globally traded hardware (Barbose and Darghouth, 2018). Utility scale systems have much more scale economies in soft costs and thus are closer to an even split between local and global costs.

The inherently local aspects—and consequent effects on employment, economic development, and interest group activity— helped stimulate policymaking in many places, particularly at the local level. Local content has provided a justification for policy adoption of demand-side solar policies (Lyon, 2016). Importantly, solar does not need local content requirements, for example in contrast to wind (Bougette and Charlier, 2015). Solar's local characteristics are inherent in the difficult-to-globalize activities of roofers, electricians, and the knowledge of local markets they rely on for sales. The local content of solar, which has only grown as the internationally traded part has shrunk, has helped maintain local support in the context of a footloose global module industry. The immobile aspects of PV installation helped buffer the effects of globalization thus enhancing local policy support. Local content has also promoted bottom-up policymaking and with all that brings both innovative designs, as well as a chaotic patchwork of policies.

Benefits of volatility

This combination of policy volatility and the shuffling of leadership on solar from one jurisdiction to another seems chaotic. It is raised as one of the main explanations for why cheap solar took so long to arrive. But it is important to understand that there were benefits to this turbulence. Foremost, this reshuffling enabled governments to learn from the experience of others. One could see what worked and what didn't. One could use policy components that were effective and remove those that generated adverse unintended consequences. This enabled policy recombination. It may have even helped countries design entirely novel components and, occasionally, completely novel policies.

By far the most important transfer of a policy innovation originated with the Public Utilities Regulatory Policy Act, a US Federal policy in 1978 that required electric utilities to buy electricity from other suppliers of electricity if those suppliers could offer it at less than the utilities' "avoided cost" (Hirsch, 1999). The level of the avoided cost of course was the most contentious aspect of the policy. It was most influentially implemented in the Standard Offer Contracts created by the California Public Utilities Commission in the 1980s. The most prominent of these was known as "Interim Standard Offer Contract #4" (ISO4) (CPUC, 1983), which gave rise to the California wind industry (Gipe, 1995). The core of the contract was that the electric utility guaranteed that it would purchase electricity at the specified purchase price for ten years. This price guarantee catalyzed billions of dollars in investment (Nemet, 2012b). The Germans looked closely at the ISO4 in designing their 1990 renewable energy law, and ultimately their groundbreaking 2000 Renewable Energy Law (EEG). It is possible that observing the experience of other countries helped to encourage an open-mindedness that led policymakers to find progress in policy more generally, for example by looking to experience outside of solar.

Wind as a precursor

Policymaking in multiple jurisdictions was central to solar's progress. It is striking how often wind power policies proceeded solar ones. For decades, wind power had been the more realistic alternative to solar. It could be installed at much larger scale. It was mostly a utility and community scale resource, not one for individual households, thus it fit the existing business models and regulatory framework better. It also had the benefit of familiarity since wind had been harnessed for ships, water pumping, and granaries for centuries; its mechanical, rather than quantum energy, made it more familiar. Above all, until recently, wind was also much more affordable than solar.

While wind has often been portrayed as a competitor to solar, it played a more influential role as a precursor technology. In effect, wind power allowed policymakers to test policies when solar was too expensive. Wind provided grid operators with experience in managing an intermittent resource on a large scale. Wind, not solar, was the main beneficiary of the ISO4 contracts in California in the 1980s. Wind was the beneficiary of the EEG in 2000; it was only in 2004 that solar adoption exploded due to the new EEG tariff rates. Renewable portfolio standard (RPS) policies in the US have boosted wind and most RPS compliance was through wind until recently. Now RPSs are one of the key drivers of demand for new solar in the US. Wind was crucial to honing policies that later became important for solar. The 2006 Chinese Renewable Energy Law focused on supporting wind power. Further, wind gave producers experience in scaling up and serving multiple national markets. Producers became familiar with policy volatility in wind policies, for example the Production Tax Credit (PTC) in the US. One benefit was the ability to hedge against policy risk by operating globally.

Hedging policy volatility

While the local aspects of solar enabled local advocacy coalitions to form and adopt policies favorable to solar, the global aspects of solar enabled equipment producers to hedge their policy risk. Solar modules are light, compact, modular, and have a relatively high value to weight ratio. They can be shipped easily around the world. The fabs that produce solar modules differ in that they are much more geographically fixed, although the valuable equipment within them is mobile. Fabs also differ in that they have to be massive to be competitive. They require hundreds of millions of dollars of investment to get started. And they take years to pay off. Consequently, expected future demand for the solar modules they produce is central to their viability, as well as to whether they are built in the first place. Scaling up production is very tightly linked to expectations of future demand for solar.

Demand for PV has been volatile in part because policies have been volatile. Solar companies face substantial policy risk in that subsidies come and go, feed-in tariff rates are reset, and prices of competing sources of electricity are spikey. Importantly, these factors are not well correlated around the world, as has been seen in wind power (Nemet, 2010). When incentives for solar are cut in one jurisdiction, there are typically other jurisdictions where they are maintained or strengthened. This provides an opportunity to hedge policy risk by operating globally and serving multiple markets. The highly globalized solar panel supply chain provides a hedge against independently uncoordinated national and subnational policies. As a CEO of one firm put it, "Sure we can make more money by selling in just that one market. But that would be too risky for us."

As discussed above, solar is distinct in that it consists of a combination of highly globalized added value (modules) and inherently local added value (installation, etc.). That share of total cost of solar electricity has shifted over time from mostly global to now mostly local. The global aspects have enabled massive economies of scale and the ability to hedge policy risk, both of which have stimulated investment. The local aspects have catalyzed public support for solar policies even in the context of a highly globalized industry.

Knowledge mobility

While policies have been a key influence in solar's evolution, individuals have also been important. Above all, the mobility of individuals—migrating, relocating, visiting, and changing companies—has been a central feature of solar's improvement. The interaction of individual mobility and policy support in the form of research and development (R&D) funding is an under-appreciated aspect of solar's technical evolution. Individuals dispersed knowledge by moving around. R&D enabled the creation of that knowledge and, crucially, preserved it when the vagaries of policy volatility threatened to scatter it.

Mobile individuals

This book focuses on the international linkages that are not evident in trade statistics or other data that could be included in regression analyses. The reason for this focus is that so much of the know-how that spread from one place to another resided in people's heads as tacit knowledge (Alic, 2008). Over the decades of solar's evolution, people were almost constantly moving around, visiting other countries, attending conferences, all of which facilitated exchange of knowledge, sometimes intentionally and often serendipitously. To be sure, hardware moved too. The modules spread easily around the world. Importantly, production equipment moved too. One interviewee estimated that $30 billion of solar manufacturing equipment moved from Germany to China between 2008 and 2012. It is important to note that in contrast to modules being transported from Asia to European markets, when production equipment moved, people came with it. Staff from equipment suppliers came and installed the equipment, and stayed to optimize the lines (Gallagher, 2014). The people are necessary because of the tacit know-how they each acquire that is then transferred. That was important, for example, when setting up and optimizing a production line, or arranging a massive solar array, or efficiently installing a small-scale system on an idiosyncratic roof. The role of material and tacit knowledge transfer is well established in many industries. The importance of knowledge transfer via professional conferences has occurred in solar since the 1950s (Taylor et al., 2007). In my interviews, I was struck by how many of the people I spoke with mentioned examples of people moving—and emphasized the importance of these visits, meetings, and experiences.

Open and observing

People and machines moved around the world and spread know-how. But a less active form of knowledge transfer—observation—was also important. People in the solar industry seemed especially open to observing what was happening in other places. Scientists read and listened to the findings of other scientists. Potential adopters, such as the US Signal Corps in the 1950s, kept close track of what was emerging from laboratories. Martin Green's group toured the US in the 1970s. Markus Real from Switzerland visited and noticed what was happening in California in the 1980s. The designers of the German feed-in tariff saw the results of the California Standard Offer Contracts. The Japanese designers of the New Sunshine Program saw Germany's first federal residential solar subsidy program, The 1000 Roofs Program. Germany's 2004 FiT inspired many, most important China's 2005 Renewable Energy Law and its 2011 feed-in tariff. Chinese solar firms rapidly adopted manufacturing innovations from each other in the post-2010 period (Ball et al., 2017). Employees would often switch to new companies over the Chinese New Year. Countries watched each other. There was a surprisingly weak inclination to be insular, to do it our way, to do it in isolation. There was instead an abundance of open-mindedness, observation, borrowing, and knowledge transfer.

Globalization enabled knowledge flows

Globalization, in its broadest sense, enabled all of this knowledge transfer. The modernization and integration of the world economy in the second half of the 20th century put in place the infrastructure on which all of this movement of knowledge depended—including tacit knowledge bound to individuals, knowledge embedded in equipment, and the knowledge codified in scientific reports. This enabling environment included transportation, which decreased in cost almost constantly, facilitating the movement of people as well as equipment. The reduction in import tariffs over time, despite recent increases, also enabled the movement of hardware. Information flows made scientific and other knowledge travel easily. Information also enabled the procurement of input materials. All of this supported cultural globalization that enabled the relocation of individuals from one country to another.

Among the first adopters of the opportunities afforded by globalization were oil companies. They had been discovering new oil reserves in far flung locations for the entire 20th century (Yergin, 1991). By the time practical solar became available, they were highly globalized. They became some of the most important early adopters of solar in the 1960s, using solar and batteries to power navigation lights on offshore oil rigs after having seen their use for buoys in Japan. They worked well. And because oil firms were global, their use spread globally quickly. Many around the world observed their effectiveness and importantly their adoption by large, global, multinational companies, who were serious about maximizing profits and delivering value to shareholders.

International innovation system

Let's briefly reconnect these observations to the academic literature. Academics call this a "global innovation system" or a series of internationally linked national innovation systems (Binz and Truffer, 2017). Culture history and institutions are central to the NIS idea. And they are absolutely important for how solar got cheap. From an innovation theory perspective, "transnational linkages" have been central to the story of solar (Wieczorek et al., 2015). One of the main ideas in the National Innovation Systems literature is that NIS are not designed, they are inherited and evolving structures.

R&D preserved knowledge

Another useful concept from the NIS literature is the "interaction of agency and institutions" (Acs et al., 2016). Much of the NIS work emphasizes the roles of broad, far-reaching institutions, such as education, and the roles of non-institutionalized influences, like culture. In solar, government-funded research and development (R&D) was an institutional mechanism that played a role and interacted importantly with "agents" in the solar industry.

The boom and bust aspect of solar described above has meant that there have been ample opportunities for knowledge to be lost. The accumulation of new knowledge is not just monotonic. Knowledge can decline in value as technology moves on and becomes obsolete (Grubler and Nemet, 2014). More perniciously, knowledge that is bound up in people's heads as tacit knowledge is vulnerable to loss if those individuals stop applying that knowledge. A knowledgeable person's retirement is a prime example. Solar's history is best seen as a series of boom-bust cycles on top of a series of relocation of leadership. As a result, retirements are not the only means by which knowledge can be lost. Companies go bankrupt and employees leave. When the industry is on a down-cycle those individuals may leave the industry completely.

Out-migration of knowledge in solar has happened several times—in the late 1950s as solar's promise gave way to other rising concerns; in the early 1980s as Carter's interest waned and then Reagan slashed budgets; in German cell production after 2008; and to some extent in China during the over-capacity binge after 2012. The volatility of solar markets has made preserving that knowledge crucial. Public R&D was central to that in that it helped codify activities, for example in reports. Many companies invested in solar R&D. The US and Japan each had R&D programs back in the 1950s and 60s. Even China, a late entrant in production, started an R&D program in the 1980s. These R&D investments established institutions that persisted even when budgets were cut and private sector investment waned.

Both public and private R&D helped with the initial discoveries (Nemet and Husmann, 2012), but government R&D's bigger role may have been to preserve knowledge in the multiple bleak periods during which people fled the industry to other sectors of the economy that were growing. Many scientists and engineers did in fact scatter to the wind to pursue the next growth industry, but public R&D kept some of them, and importantly codified much of what had been learned. This knowledge was kept alive until the next niche or next policy or when another driver of the next boom period arrived.

Technical progress

The technology of PV itself evolved from a combination of: early convergence on a dominant design; a clear theoretical foundation of the photoelectric effect; disruptive manufacturing; and implementation of policies oriented toward learning in 1975 (US), in 1994 (Japan), in 2004 (Germany), and in 2011 (China).

BOX 2.3 COMPONENTS OF A PHOTOVOLTAIC SYSTEM.

PV (photovoltaics): Semiconductor material that has the ability to generate free electrons when exposed to light.
Cell: A PV material that includes electrical contacts and coatings to generate an electric current.
Module: A set of PV cells that are wired together and encapsulated in glass and plastic. Also known as a solar "panel."
Inverter: An electronic device that converts direct current (generated from PV) into alternating current, used in electric grid and households.
Racking: Support structures used to mount PV modules on the ground or on a roof.
Hardware: Modules, inverters, and sometimes including racking equipment.
PV system: Hardware as well as wiring used to connect a PV module with a local electrical system.

An early dominant design

One of the surprising attributes of solar's evolution is that it still uses the same material and general design of the technology from 1954. Crystallized silicon has maintained over a 90% share of PV production, despite a wide array of different materials and configurations. The concept from the innovation literature on "dominant designs" provides another insight on why solar has been so successful (Suarez and Utterback, 1995). The idea is that technologies are amenable to scale up once a dominant design is established, which is defined as "stable core components that can be stable interfaces" (Murmann and Frenken, 2005). Solar's overall design has been remarkably stable—in part since 1954 and especially since 1985—despite the emergence of many competing materials, designs, and combinations with at least some attributes that are superior.

Repeatedly, silicon has held off competitors—even in the 2007–08 period when silicon prices spiked. The continued room for improvement, when so many times it has been declared a dead-end, echoes the Moore's Law in computing (Farmer and Lafond, 2016). Silicon-based microchips have similarly induced hand-wringing about the end of 50 years of increases in the density of processing power, but which still has not arrived (Nagy et al., 2011). Leading companies in Japan, Germany, and the US lost their solar business in part by investing in alternatives to silicon. As one interviewee put it: "never bet against silicon." One of the reasons is due to silicon's pervasive use in the computer industry and the accumulation of know-how that has resulted. There is ample knowledge in computers that has been borrowed and repurposed for solar. The industry's scrap provided a source of silicon for a long time until PV needs exceeded scrap supply from computers. But expertise in scale up was also an important benefit of this inter-industry spillover (Nemet, 2012a).

The current dominance of silicon does not mean that it is entrenched forever. Indeed, alternatives such as perovskite, organic, and quantum dot solar cells are all improving in efficiency and reliability (Sivaram, 2018). They may be set up to take on solar, potentially at even lower costs than silicon, albeit once they scale up. But the cheapness of silicon makes the barriers to entry—think about a billion-dollar fabrication facility, or fab—much more substantial than in the past. Hybrids, such as perovskites layered on top of silicon seem more likely, as they can piggyback on the scale economies and supply chain established by silicon.

Theoretical basis

One should also note scientists' knowledge creation and knowledge mobility depended on a fundamental understanding of the physics at work. That all derived from over one hundred years ago and Einstein's theory describing the photoelectric effect as photons, rather than waves (Einstein, 1905). That understanding was critical to the work at Bell Labs designing the p–n junction for the first efficient cell (Perlin, 2013). The link between Einstein and Bell fits well with a theory of invention arguing that key inventions require some theoretical understanding of a phenomenon in order to successfully exploit it (Arthur, 2009). This does not support the notion we just need more science and R&D (Bush, 1945; Lomborg, 2010). PV did not begin with Einstein—there were over 60 years of experiments with strange responses of materials' electrical properties when exposed to light. R&D certainly did not get us to today's $18 per MWh PV by itself.

Disruptive manufacturing

One thing that did take us there was how PV was produced. PV manufacturing was a classic disruptive technology story (Wilson, 2018). Most adopters did not need the high efficiency and low degradation rates that the space program applications required. In the 2000s, Chinese technology was not as good as the German, but it was good enough to make electricity on Earth, even if not good enough for space. The Chinese compromised on the right attributes while still obtaining international certification and made reliable cells with high-enough efficiency. At the same time, the Chinese found channels to raise the credibility of their initial products, such as partnering with installers in Germany who were desperate to meet surging demand in 2008–11. Then they gradually upgraded their reliability via technology standards. But a key contribution was to relax some of the exacting tolerances that had previously made production slow and expensive.

R&D is insufficient

The disruptive manufacturing hypothesis illustrates the point that much improvement in solar happened outside of R&D funding. R&D was crucial for technical improvements, but it was far from sufficient to take solar from the initial Bell Labs' device to the massive industry we have today. That, more than anything, is a lesson for other technologies. Perhaps the most compelling depiction of PV's progress has been the "learning curve." Paul Maycock, the first director of the US Department of Energy's (called ERDA at the time) solar program had become familiar with learning curves while working at Texas Instruments (TI) in the early 1970s. TI may have picked it up from the Boston Consulting Group who popularized it at the time (BCG, 1972). There is evidence of learning curve thinking in PV in the 1950s (Cherry, 1955). In 1975, Maycock plotted the cost of PV as a function of experience (Maycock and Wakefield, 1975) (see Chapter 4) and that concept has been used in PV ever since. It has had a clear policy implication: R&D is insufficient. Policymakers could "buy down" the cost of PV in the same way that Texas Instruments' electronic products had used "forward pricing"—take losses early to scale up manufacturing and achieve cost reductions and profits later. The "Block Buy" programs from 1975 to 1985 implemented this concept (Christensen, 1985). The learning curve was also a justification for the over $200 billion spent on PV subsidies in Germany (Unnerstal, 2017). Remarkably, over forty years later we are still on Maycock's figure. By 2018 the industry had accumulated 500 GW of production volume and cells were selling for $0.20 per W, which is $0.06 per W in 1975 dollars, or $56 per kW. PV has traveled down the most optimistic of Maycock's learning curve projections, as we are now at the far edge of his foresight.

A long development path

Even though the emergence of cheap PV over the past few years seems dramatic it has not been sudden. It has been over a century since Einstein's theory of the photoelectric effect. We were able to ride the most optimistic version of Paul Maycock's 1975 learning curve, but that took us 42 years. Once one accepts that solar is in fact now astonishingly cheap, one could ask not only, how did solar get cheap? But also: *why did it take so long?* If we want to use PV as a model for other emerging climate technologies, we need to understand the full technology development trajectory, not just the past 15 years of explosive growth. Understanding PV's "formative phase" (Jacobsson and Bergek, 2004) is the primary justification for a careful look at the important dynamics of PV's development before 1973.

References

Acs, Z. J., Audretsch, D. B., Lehmann, E. E. & Licht, G. 2016. National systems of innovation. *The Journal of Technology Transfer*, 42, 997–1008.

Alic, J. A. 2008. Technical knowledge and experiential learning: What people know and can do. *Technology Analysis & Strategic Management*, 20, 427–442.

Arthur, W. B. 2009. *The Nature of Technology: What It Is and How It Evolves*, New York, Free Press.

Ball, J., Reicher, D., Sun, X. & Pollock, C. 2017. *The New Solar System: China's Evolving Solar Industry and Its Implications for Competitive Solar Power in the United States and the World*, Steyer-Taylor Center for Energy Policy and Finance.

Barbose, G. L., Darghouth, N. R., Millstein, D., Lacommare, K. H., Disanti, N. & Widiss, R. 2017. *Tracking the Sun 10: The Installed Price of Residential and Non-Residential Photovoltaic Systems in the United States*, Lawrence Berkeley National Laboratory and SunShot.

Barbose, G. L. & Darghouth, N. R. 2018. *Tracking the Sun: Installed Price Trends for Distriuted Photovoltaic Systems in the United States*. Berkeley, CA, Lawrence Berkeley National Laboratory.

BCG 1972. *Perspectives on Experience*, Boston, The Boston Consulting Group.

Binz, C. & Truffer, B. 2017. Global Innovation Systems—A conceptual framework for innovation dynamics in transnational contexts. *Research Policy*, 46, 1284–1298.

Bolinger, M., Seel, J. & Lacommare, K. H. 2017. *Utility-Scale Solar 2016: An Empirical Analysis of Project Cost, Performance, and Pricing Trends in the United States*. Berkeley, CA, Lawrence Berkeley National Lab. (LBNL).

Bougette, P. & Charlier, C. 2015. Renewable energy, subsidies, and the WTO: Where has the 'green' gone? *Energy Economics*, 51, 407–416.

Bush, V. 1945. *Science The Endless Frontier: A Report to the President by Vannevar Bush, Director of the Office of Scientific Research and Development*. Washington, DC, United States Government Printing Office.

Cherry, W. 1955. Military considerations for a photovoltaic solar energy converter. Transactions of the International Conference on the Use of Solar Energy, Tucson, Arizona, 31 October – 1 November 1955.

Christensen, E. 1985. *Electricity from Photovoltaic Solar Cells: Flat-Plate Solar Array Project of the U.S. Department of Energy's National Photovoltaics Program, 10 years of progress. JPL publication: 400–279*, Pasadena, CA, Prepared by the Jet Propulsion Laboratory for the U.S. Dept. of Energy through an agreement with the National Aeronautics and Space Administration.

CPUC 1983. *Approval of Interim Standard Offer No.4 for PG&E, SCE, and SDG&E based on the results of a Negotiating Conference, Decision 83–09–054*. California Public Utilities Commission.

Davidsson, S. & Höök, M. 2017. Material requirements and availability for multi-terawatt deployment of photovoltaics. *Energy Policy*, 108, 574–582.

Dunford, M., Lee, K. H., Liu, W. & Yeung, G. 2013. Geographical interdependence, international trade and economic dynamics: The Chinese and German solar energy industries. *European Urban and Regional Studies*, 20, 14–36.

Einstein, A. 1905. Über einen die Erzeugung und Verwandlung des Lichtes betreffenden heuristischen Gesichtspunkt. *Annalen der Physik*, 322, 132–148.

Farmer, J. D. & Lafond, F. 2016. How predictable is technological progress? *Research Policy*, 45, 647–665.

Friedman, B., Ardani, K., Feldman, D., Citron, R., Margolis, R. & Zuboy, J. 2013. *Benchmarking Non-Hardware Balance-of-System (Soft) Costs for US Photovoltaic Systems Using a Bottom-Up Approach and Installer Survey*. NREL.

Gallagher, K. S. 2014. *The Globalization of Clean Energy Technology: Lessons from China*, Cambridge, MA, MIT Press.

Gipe, P. 1995. *Wind Energy Comes of Age*, New York, Wiley.

Goodrich, A. C., Powell, D. M., James, T. L., Woodhouse, M. & Buonassisi, T. 2013. Assessing the drivers of regional trends in solar photovoltaic manufacturing. *Energy & Environmental Science*, 6, 2811–2821.

Grubler, A. & Nemet, G. F. 2014. Sources and consequences of knowledge depreciation. In: Grubler, A. & Wilson, C. (eds.) *Energy Technology Innovation: Learning from Historical Successes and Failures.* Cambridge, Cambridge University Press.

Haegel, N. M., Margolis, R., Buonassisi, T., Feldman, D., Froitzheim, A., Garabedian, R., Green, M., Glunz, S., Henning, H.-M., Holder, B., Kaizuka, I., Kroposki, B., Matsubara, K., Niki, S., Sakurai, K., Schindler, R. A., Tumas, W., Weber, E. R., Wilson, G., Woodhouse, M. & Kurtz, S. 2017. Terawatt-scale photovoltaics: Trajectories and challenges. *Science,* 356, 141–143.

Hirsch, R. F. 1999. *Power Loss: The Origins of Deregulation and Restructuring in the American Electric Utility System,* Cambridge, MA, The MIT Press.

Jacobsson, S. & Bergek, A. 2004. Transforming the energy sector: The evolution of technological systems in renewable energy technology. *Industrial and Corporate Change,* 13, 815–849.

Jenkins-Smith, H., Nohrstedt, D., Weible, C. & Sabatier, P. 2014. The Advocacy Coalition Framework: Foundations, evolution, and ongoing research. In: Sabatier, P. & Weible, C. (eds.) *Theories of the Policy Process.* Boulder, CO, Westview Press.

Karakaya, E., Nuur, C. & Hidalgo, A. 2016. Business model challenge: Lessons from a local solar company. *Renewable Energy,* 85, 1026–1035.

Kemp, R., Schot, J. & Hoogma, R. 1998. Regime shifts to sustainability through processes of niche formation: The approach of strategic niche management. *Technology Analysis & Strategic Management,* 10, 175–198.

Kingdon, J. W. 1984. *Agendas, Alternatives, and Public Policies,* Boston, Little, Brown.

Kivimaa, P. & Kern, F. 2015. Creative destruction or mere niche support? Innovation policy mixes for sustainability transitions. *Research Policy,* 45, 205–217.

Laird, F. N. 2003. Constructing the future: Advocating energy technologies in the Cold War. *Technology and Culture,* 44, 27–49.

Lomborg, B. 2010. *Cool It: The Skeptical Environmentalist's Guide to Global Warming,* Vintage.

Lyon, T. P. 2016. Drivers and impacts of renewable portfolio standards. *Annual Review of Resource Economics,* 8, 141–155.

Maycock, P. D. & Wakefield, G. F. 1975. Business Analysis of Solar Photovoltaic Energy Conversion. 11th IEEE Photovoltaic Specialists Conference, May 6–8 1975 New York. IEEE, 252–255.

Murmann, J. P. & Frenken, K. 2005. New Directions in Research on Dominant Designs. Academy of Management, August 2005 Honolulu.

Nagy, B., Farmer, J. D., Trancik, J. E. & Gonzales, J. P. 2011. Superexponential long-term trends in information technology. *Technological Forecasting and Social Change,* 78, 1356–1364.

Nemet, G. F. 2006. Beyond the learning curve: Factors influencing cost reductions in photovoltaics. *Energy Policy,* 34, 3218–3232.

Nemet, G. F. 2010. Robust incentives and the design of a climate change governance regime. *Energy Policy,* 38, 7216–7225.

Nemet, G. F. 2012a. Inter-technology knowledge spillovers for energy technologies. *Energy Economics,* 34, 1259–1270.

Nemet, G. F. 2012b. Subsidies for new technologies and knowledge spillovers from learning by doing. *Journal of Policy Analysis and Management,* 31, 601–622.

Nemet, G. F. & Kammen, D. M. 2007. U.S. energy research and development: Declining investment, increasing need, and the feasibility of expansion. *Energy Policy,* 35, 746–755.

Nemet, G. & Husmann, D. 2012. Historical and future cost dynamics of Photovoltaic technology. In: Sayigh, A. (ed.) *Comprehensive Renewable Energy.* Oxford, Elsevier.

Nemet, G. F., Braden, P., Cubero, E. & Rimal, B. 2014. Four decades of multiyear targets in energy policy: Aspirations or credible commitments? *Wiley Interdisciplinary Reviews: Energy and Environment*, 3, 522–533.

Nemet, G. F., Grubler, A. & Kammen, D. 2016. Countercyclical energy and climate policy for the U.S. *Wiley Interdisciplinary Reviews: Climate Change*, 7, 5–12.

NYTimes. 1954. Vast power of the sun is tapped by battery using sand ingredient. *New York Times*, p. 1.

Perlin, J. 2013. *Let it Shine: The 6,000-year Story of Solar Energy*, New World Library.

President's Materials Policy Commission 1952. *Resources for Freedom*. Washington, DC, Govt. Printing Office.

Raven, R., Kern, F., Verhees, B. & Smith, A. 2016. Niche construction and empowerment through socio-political work. A meta-analysis of six low-carbon technology cases. *Environmental Innovation and Societal Transitions*, 18, 164–180.

Schuelke-Leech, B.-A. 2014. Volatility in federal funding of energy R&D. *Energy Policy*, 67, 943–950.

Sivaram, V. 2018. *Taming the Sun: Innovations to Harness Solar Energy and Power the Planet*, Cambridge, MA, The MIT Press.

Suarez, F. F. & Utterback, J. M. 1995. Dominant designs and the survival of firms. *Strategic Management Journal*, 16, 415–430.

Tatsuta, M. 1996. New sunshine project and new trend of PV R&D program in Japan. *Renewable Energy*, 8, 40–43.

Taylor, M., Nemet, G., Colvin, M., Begley, L., Wadia, C. & Dillavou, T. 2007. *Government Actions and Innovation in Clean Energy Technologies: The Cases of Photovoltaic Cells, Solar Thermal Electric Power, and Solar Water Heating, CEC-500-2007-012*. Sacramento, California Energy Commission.

United States Department Of Energy 1979. *Domestic Policy Review of Solar Energy*, U.S. Department of Energy.

Unnerstal, T. 2017. *The German Energy Transition: Design, Implementation, Costs, and Lessons*, Springer.

Verdolini, E., Anadon, L. D., Lu, J. & Nemet, G. F. 2015. The effects of expert selection, elicitation design, and R&D assumptions on experts' estimates of the future costs of photovoltaics. *Energy Policy*, 80, 233–243.

Wieczorek, A. J., Raven, R. & Berkhout, F. 2015. Transnational linkages in sustainability experiments: A typology and the case of solar photovoltaic energy in India. *Environmental Innovation and Societal Transitions*, 17, 149–165.

Wilson, C. 2018. Disruptive low-carbon innovations. *Energy Research & Social Science*, 37, 216–223.

Yergin, D. 1991. *The Prize: The Epic Quest for Oil, Money, and Power*, New York, Simon and Schuster.

PART 1
Creating a technology

3
SCIENTIFIC ORIGINS

One of PV's most intriguing characteristics is its intrinsic connection to once cutting edge scientific concepts—such as the structure of atoms, the characteristics of light, and their interactions with each other. That connection to advanced science makes it similar to nuclear power. The early activity in solar PV was closely tied to basic science and fundamental understanding, with barely any consideration about its potential applications. Progress in this early period was extremely slow. It took 115 years from Becquerel's discovery of the photoelectric effect in 1839 until Bell Labs' demonstration of the first efficient solar cell in 1954. For the first half of that period the early tinkerers were trying to observe what they were seeing and then to replicate and enhance it. While the observations themselves were important, none made any serious progress in trying to understand these observations until Philip Lenard in the 1890s. But it was the giant of 20th-century science, Albert Einstein, who provided a full explanation of the photoelectric effect in 1905, for which he won the Nobel Prize in 1921. Einstein's work explained the phenomenon which had been observed for more than sixty years. He described light as packets of waves, or photons, which release electrons in a photosensitive material when those photons exceed a threshold energy corresponding to a material's band-gap. That understanding opened a vein of research for other scientists, like Millikan, and led to systematic development of the technology in the 1930s, first by Siemens and later by Bell Labs. Bell Labs' breakthrough in 1954 launched the PV industry and Sputnik in 1957 spurred US entry into the space race. The most enthusiastic early adopter was the US Defense department for its space program in the 1950s and 60s. The other significant early adopters were international oil companies that used PV to light offshore oil rigs. An array of niche markets emerged, albeit all of modest size. By the early 1970s solar had been established as a reliable technology and was positioned for a more significant role when the first oil crisis hit in 1973.

Scientists

The development of solar PV originates in 1839 when Edmond Becquerel observed the photoelectric effect while immersing silver chloride in an acid and exposing it to light. Even though he was only 19, this discovery was not an accident. Becquerel was working in his father's laboratory in Paris and had connected the silver chloride to platinum electrical terminals to see if a current could be observed. Becquerel's name is more famous due to his son Henri's work with radioactivity, after which the standard unit of radioactivity, the Becquerel (Bq), one decay per second, was named.

After Becquerel's discovery, progress moved to England. In 1949, inventor Alfred Smee first referred to it as the "photovoltaic effect" and in 1873 an English engineer, Willoughby Smith, observed the effect using rods of selenium while developing materials for underwater cables. Selenium was later used for the first Bell Labs PV cells in the 1950s. William Grylls Adams and R. Evans Day were the first team to systematically investigate the photovoltaic effect, publishing a paper for the Royal Society on their observations of selenium to examine "whether light could actually generate an electric current in selenium" (Adams and Day, 1877).

Charles Fritts in New York developed the first PV module in 1884, again using selenium. In line with the zeitgeist of invention and possibility in 1880s New York, Fritts spoke of solar cells as competitors to Thomas Edison's coal fired power plants, which had just come on line earlier that decade. Fritts showed his cell to Werner von Siemens, founder of the conglomerate and a prominent engineer, who brought it back to Germany and promoted its possibilities there. Around the same time, other Germans were active in making observations in this area. Heinrich Hertz found that ultraviolet light affects the sparking voltage of a gas and Wilhelm Hallwachs found that copper could also be light sensitive. With these observations and experiments in place by 1890, the world had a working PV device, albeit with less than 1% efficiency, well before they understood why it produced a current and how it worked. Siemens pushed his scientists to study this effect and understand it so society could benefit from its potential.

Einstein

At the end of the 19th century, a handful of scientists began to try to understand how light could generate an electric current. George Minchin worked with selenium and built sensitive photoelectric devices. His experiments with this more precise equipment led him to suggest that characteristics of the absorbing material affected how much electrical energy was created, which he used to posit that the wavelength of the light could play a role. Indeed, Philip Lenard studied the photoelectric effect in the 1890s and found that the electrical output was sensitive to the wavelength of the light used.

In his annus mirabilis, 1905, Einstein published (translated from German) "On a Heuristic Viewpoint Concerning the Production and Transformation of Light"

(Einstein, 1905). In that paper Einstein theorized that light consists of discrete packets of energy, light quanta, or photons. The various wavelengths in which light comes determine how much energy they contain. If the photons contain enough energy to exceed a material's band-gap, or activation energy, they will dislodge an electron, which is why experimenters were detecting a current in their photosensitive devices. In 1909, Robert Millikan confirmed Einstein's hypothesis by conducting an experiment in which his lab measured the electric charge from a tiny droplet of oil. From that result he calculated an estimate for the charge on a single electron (Millikan, 1913). More broadly, Millikan confirmed Einstein's theory about the quantum nature of light.

Einstein's theory of photoelectric was important to the subsequent development of a practical commercial solar device. As economist and technology historian Brian Arthur puts it,

> Technologies issue forth as knowledge of phenomena and their theory is gained, so that novel technologies emerge both from the cumulation of existing building-block technologies and from understandings of the phenomena that surround these. (Arthur, 2007)

In this case, the phenomenon was the photoelectric effect. Key to understanding this effect was the notion of a material's band gap and that light's activation energy could liberate electrons. This new phenomenon helped technology developers come up with the p-n junction. Once Einstein's theory was laid out and Millikan confirmed it experimentally, the door was open for scientists at Siemens, and later at Bell Labs, to work out a device that could be used not just for advancing science, but for providing practical benefits. Siemens focused on copper-based solar in the 1930s and invented a 1% efficient cell. The industrial conglomerate had ambitious plans to produce a massive amount of PV cells to deploy in the Sahara Desert and wire power to Europe. The world-leading activity in PV device technology was occurring in Germany in the 1930s and would likely have proceeded to commercial devices were it not for the Second World War. Instead Bell Labs— full of scientists, emphasizing combinations of diverse expertise, and experienced with both selenium and silicon—made full use of the understanding about photons and electrons that Einstein's theoretical structure provided.

Bell Labs

The first breakthroughs in developing a practical solar cell took place at an office park in Murray Hill, New Jersey. Founded in 1925 as Bell Telephone Laboratories, Inc., Bell Labs served the dual purpose of developing new technologies that could be used by the parent company's communications business while also investing its profits in socially beneficial applications (Gertner, 2013). The latter was an important consideration since Bell Telephone had a national monopoly on telephone services and thus, needed to maintain its social license to operate by showing the societal benefits it produced to keep regulators at bay. On paper, Bell Labs served as the R&D arm of Bell Telephone, but in practice the men and women working in its research facilities had broad freedom to pursue intriguing new technologies. Indeed, the Lab Director during the 1950s, Mervin Kelly, referred to Bell Labs as an "institute of creative technology."

Kelly was born in rural Missouri and studied physics at the University of Chicago. He rose through the Bell Labs' hierarchy quickly, serving as director of research, vice president, and eventually president of Bell Labs from 1951 to 1959. Kelly created the mantra that Bell was "inventing ways to invent things" and viewed a central part of his role as developing new ways to perform research. One means of doing so was to help design the Bell Labs building in Murray Hill, New Jersey, a space that promoted collaboration between researchers by placing them physically close to one another. His vision was to create a critical mass of talented people to generate a fast and healthy exchange of ideas (Bell Labs, 2018). He was especially interested in combining a diversity of backgrounds—so that differing domains of expertise might complement each other and combine in novel and sometimes unexpected ways. His approach was successful in that, in his time as Lab Director, Bell produced various world-changing products such as the transatlantic telephone cable, the transistor computer, the laser, and the world's first practical solar cell.

While scientists at Bell Labs had intermittently explored the science behind the photoelectric effect during the 1930s, they began to focus on it in February 1940. Engineer Russel Shoemaker Ohl had been investigating a way to use crystals in radio transmission when he noticed that one of his high-purity silicon crystals facilitated a flow of electrical current when exposed to light (Chodos, 2009). The crack in the crystal served as a boundary between positively and negatively charged impurities within the crystal, a structure later called the "p–n junction." Because part of the crystal was contaminated with phosphorus, it had extra electrons and formed a negative or n-type (negative) semi-conductor. The other side of the junction contained boron, which meant it was missing electrons and thus represented a p-type (positive) semi-conductor. Ohl had discovered "functional impurity." He showed the effect to some scientist friends, including Lab Director Kelly. In the patent application he filed in May 1941, Ohl described the discovery,

> During an investigation of the production of fused silicon of high purity and its use for point contact rectifiers, the applicant discovered that under certain conditions this material was sensitive to visible light, generating an electromotive force independently of any applied voltage. The light sensitive effects were of a magnitude comparable to the most effective photoelectric substances then known. (Ohl, 1946)

It took 14 years before Ohl's discovery became developed as the first "solar cell."

That project began when Calvin Fuller, Gerald Pearson, and Daryl Chapin came together to form the "solar battery project," almost purely by chance. Before solar photovoltaic energy became an emphasis in the mid-1950s, many at Bell Labs were working on the transistor. This project was born out of a wartime effort to create highly purified germanium for use in radar systems. While the transistor emerged from this research program, so did a cheaper and more durable material to replace germanium, silicon. Bell's transistors revolutionized electronics and provided the basis for integrated circuits and microprocessors (systems of transistors), which form the core of computing equipment. Silicon also turned out to be a good material for producing solar PV cells.

Chapin, an electrical engineer, began tinkering with selenium solar cells in early 1952 as a means of powering a telephone service in remote areas of Latin America. However, the selenium cells were extremely inefficient. His friend, Gerald Pearson, an experimental physicist as well as one of the leading scientists on the silicon transistor and expert on silicon, recommended that Chapin replace the selenium with silicon. Their goals were to develop better transistors, which because they required little power, could be useful in remote applications. Thus, the Bell Solar Battery Project began and was able to piggyback on the successes of the transistor project—each improvement to the silicon for transistors also led to better efficiency for the silicon solar cell. The breakthrough was the doping of silicon, i.e. the introduction of impurities into the silicon surface, to manipulate its electrical properties. Calvin Fuller, a chemical engineer and the third member of the solar battery team, led the doping process as well as preparing highly pure silicon. By doping two pieces of silicon with different impurities, and then joining the surfaces, he intentionally created the p–n junction that Ohl had discovered a decade earlier.

Chapin, Fuller, and Pearson tried a wide array of configurations to reach their goal of 6% efficiency. Each time they hit a wall stymieing them from further efficiency gains, Fuller was able to produce what solar history expert John Perlin calls "the sauce," higher purity silicon, more uniform doping, and better junctions (Perlin, 1999). Morton Prince was brought on in February 1954 to optimize the device and make it reliable enough for a demonstration (Palz, 2010). In March, they reached their 6% efficiency goal.

Indicative of solar's progress, they did not hide the breakthrough but instead immediately published it in an academic journal, in which they described both the 6% achievement and the reasons why they fell far short of the theoretical limit of 22%, such as reflectance and recombination (Chapin et al., 1954). On March 5, 1954 the three Bell scientists filed a patent application for a "Solar Energy Converting" apparatus (Chapin et al., 1957). The patent cited prior art from eight previous patents—the multiple technologies that were brought together to form the solar cell (Figure 3.1). They included the first patent by Ohl, as well as one for his "improved" light sensitive device (Ohl, 1948). Also cited were three other material processing patents for producing high purity silicon (Scaff, 1946), semiconductor crystals (Kirkpatrick and Sears, 1952), and p–n junctions (Sparks, 1953). The other three were electronics patents, for photoelectromotive force (Teal, 1947), direct current reverse power controller (Bell, 1952), and a variable impedance device (Barton, 1951). The first six were all filed by inventors working at

60 Creating a technology

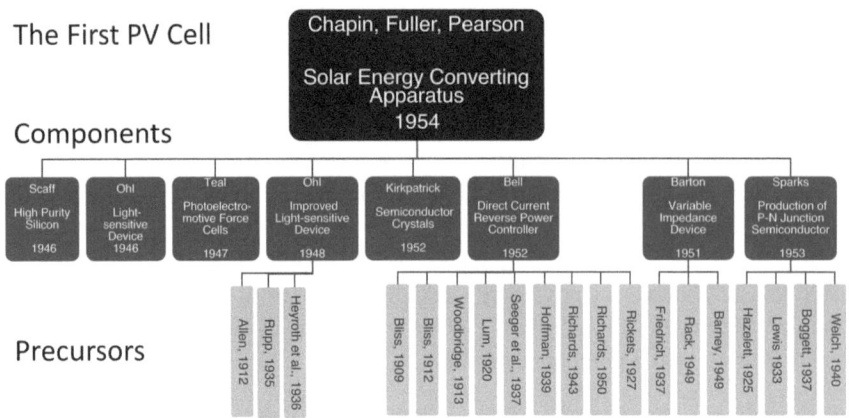

FIGURE 3.1 Prior art leading to the Bell Labs patent for a Solar Energy Converting Apparatus.

Bell Labs. The reverse power controller was by Emerson Bell, an independent inventor, and the variable impedance device was from RCA. While the effectiveness of the p–n junction depended on the notion of activation energy from Einstein, of the 1063 papers that cite the Einstein 1905 paper, only two were authored by Bell Labs' scientists–in the 1970s and 80s. Moreover, none of the Bell patents cite Einstein, either; Bell scientists seemed much more likely to cite other Bell scientists.

On April 25, 1954, Bell Labs' managers held a press conference announcing the first solar battery, using room light to power a miniature Ferris wheel. The Labs' vision for solar was grandiose and they promoted the invention in national magazine advertisements. Expectations were high and the announcement received front-page coverage in the *New York Times*, which echoed the Labs' ambition for solar:

> It may mark the beginning of a new era, leading eventually to the realization of one of mankind's most cherished dreams—the harnessing of the almost limitless energy of the sun for the uses of civilization. (NYTimes, 1954)

Indeed, the first paper making reference to a crude solar PV learning curve was presented at a conference the following year (Cherry, 1955). Prince identified opportunities for improvement (Prince, 1955) and by mid-1955, the group had reached 9% efficiency. While this event was certainly the most important breakthrough in the history of PV, it is humbling to note that it was not quick to arrive—it took 117 years from Becquerel's discovery in 1837.

Bell first put the invention to use with a 10 Watt module at a test telephony application in Americus, Georgia in late 1955. The cells worked but were prohibitively expensive to produce. In fact, in 1956, Chapin calculated that it would cost $1.5 million to power a home, clearly failing to meet the "better or cheaper or both" dictum of Bell Labs. Still, the cells were a technical success to the extent that most of the what the team of three put together under the Solar Battery Project remains essential to the design of photovoltaics today. Crucially, silicon emerged as the key material. Other materials had better properties, for example the band gap of some more directly corresponds to the wavelengths found in sunlight. But silicon is a stable element and just as important, there were increasing returns to using it. As people worked with it, not just for solar but also for computer applications, they became familiar with its properties and figured out how to make it a more effective converter of light to electricity—and there turned out to be plenty of room for improvement. Indeed, cost reductions in crystallized silicon continue to be found more than sixty years later.

An emerging commercial market

Bell Labs' invention and promotion of solar, launched an industry. A set of actors emerged interested in pursuing the opportunities that solar presented. The earliest of these early adopters was the US military.

After the Second World War, the US government had recruited many German engineers and scientists under Operation Paperclip as a means of gaining an advantage over the Soviet Union in the both the space race and the arms races of the ensuing Cold War. One such scientist, Hans Zeigler, was a pioneer of communications satellite technology and led efforts to incorporate solar as their main power source. Upon his arrival in the United States, Zeigler began working for the US Army Signal Corps, which saw the potential of solar for remote and emergency applications in its communications technology. Ziegler visited Bell Labs in 1955 with the idea of solar for satellites in mind, which began as "Operation Lunch Box" (Perlin, 1999).

The military's interest came on top of a gathering concern about energy resources after the Second World War. Energy scarcity played a role in many of the conflicts, such as Rommel's struggles in North Africa, Japan's invasion of Indonesia, and the Nazis' scale up of coal-to-liquids plants. The UN Scientific Conference on the Conservation and Utilization of Resources in New York in 1949 highlighted energy as one of the critical resource scarcity concerns for the world. In 1952, the Paley Commission appointed by the Eisenhower administration warned of increasing dependence on Middle Eastern oil and proposed R&D on solar (Paley Commission, 1952). The Association for Applied Solar Energy was established in 1954 after a symposium at the University of Wisconsin in 1953.

Still, solar seemed to be a disappointment in the immediate aftermath of Bell's 1954 announcement. As it turned out, 1954 was also the year of groundbreaking for the first commercial nuclear reactor at Shippingport, PA. Eisenhower had made his "Atoms for Peace" speech in December 1953, touting the civilian benefits of nuclear technology both in the US and as exports to other countries such as Israel, Iran, and Pakistan. Nuclear power became the dominant emerging energy technology of the 1950s and solar technology was losing support and was close to being put back on the shelf.

Solar's status changed quickly with the Soviet launch of the first Earth orbiting satellite, Sputnik, in October 1957. Suddenly the US seemed to have fallen behind in the space race and resources quickly became available to elevate the US effort. The US Navy won the competition against the Air Force and Army to lead the satellite program. The Navy wanted to use batteries, a proven technology, to power the satellite. Ziegler pushed them to consider solar, arguing that solar cells would last longer than batteries and that they were lighter. Indeed, the first US Satellite, Explorer-I used mercury batteries that accounted for 40% of its weight and lasted only four months. The Navy took a chance on solar for the Vanguard-1 satellite launched in March 1958. Its six solar cells powered the radio transmitter for six years. Vanguard-1 demonstrated that solar was a serious technology and from 1958 to 1969, the US space program spent approximately $50 million on ten million solar cells, in addition to another $100 million for glass and balance of system components (NRC, 1972). By 1972, about a thousand spacecraft had been launched using solar cells.

The satellite market was a tremendous boost for the solar industry and firms entered the industry to meet the demand. The first mover, National Fabricated Products (NFP) in Chicago, bought the licensing rights to the Bell solar patent from Western Electric and began shipping the first commercial solar cell, the S-400, in June 1955. Hoffman Electronics, founded in California by Les Hoffman to make televisions, bought NFP in 1956 and targeted the cells for telecom repeater stations. Hoffman had experience producing electronic equipment for the military and had developed a reputation with them for reliability. In 1956, Hoffman hired Morton Prince who had optimized the Bell Labs cell in the lead up to the 1954 demonstration. Hoffman scientists, led by Prince as director of R&D, improved the efficiency of these early commercial cells, for example introducing grid-type electrical contacts, and produced the six cells that were used on Vanguard-1 (Nemet and Husmann, 2012). That success led to orders of thousands of cells. By the end of the 1950s, Hoffman began marketing a solar powered radio, the first consumer product using solar cells. The company was also involved in the design and implementation of the world's first solar powered, coast-to-coast, two-way, broadcast transmission by the US Army Signal Corps (Perlin, 1999).

New niche markets and new companies emerged in the 1960s. Navigational lighting applications began in Japan with cells produced by Sharp. Elliot Berman of Exxon observed these systems on a trip to Japan and made the case for solar within Exxon. Solar navigation lighting was soon adopted by the US Coast Guard, and by oil companies that had developed off-shore drilling platforms in the Gulf of Mexico. Tideland Signal, Exxon, and other oil and gas companies became customers and Automatic Power, Sharp, Philips, and Solar Power Corporation began to serve that market. Because many of these customers were international, solar technology diffused to multiple countries in the 1960s. In many cases, such as remote railway crossings and telecom repeater stations, solar powered batteries were more reliable than grid electricity. Developing country applications, such as solar powered water pumps installed by the French in Mali further diffused the technology.

Harbingers of a new urgency

These applications demonstrated that solar was more than an eccentric contraption of interest to physicists and tinkerers. It had real world applications in a variety of domains. Demanding customers, such as the US military and international oil companies, were using them over extended periods of time. They saw that they were reliable and affordable in situations where the cost of other forms of power was high, such as in space and in remote locations. And thus, an industry had emerged to produce solar cells. It was not highly automated and still small scale. Yet, they were producing products that worked well and were reliable over many years with minimal maintenance. As the 1970s began, these were promising signs that perhaps solar would be up to the challenge of serving much larger markets and becoming a competitor to fossil fuel power.

References

Adams, W. G. & Day, R. E. 1877. The action of light on selenium. *Proceedings of the Royal Society of London*, 25, 113–117.
Arthur, W. B. 2007. The structure of invention. *Research Policy*, 36, 274–287.
Barton, L. E. 1951. *Variable impedance device*.
Bell, E. D. W. 1952. *Direct current reverse power controller*.
Bell Labs. 2018. *From AT&T to Nokia, a continuous focus on industry innovation and scientific exploration* [Online]. Available: https://www.bell-labs.com/explore/history-bell-labs/ [Accessed January 15 2018].
Chapin, D. M., Fuller, C. S. & Pearson, G. L. 1954. A new silicon p–n junction photocell for converting solar radiation into electrical power . *Journal of Applied Physics*, 25, 676–677.
Chapin, D. M., Fuller, C. S. & Pearson, G. L. 1957. *Solar energy converting apparatus*.
Cherry, W. 1955. Military considerations for a photovoltaic solar energy converter. Transactions of the International Conference on the Use of Solar Energy, Tucson, Arizona, 31 October – 1 November 1955.
Chodos, A. 2009. Bell Labs demonstrates the first practical silicon solar cell. *APS News*, 18.
Einstein, A. 1905. Über einen die Erzeugung und Verwandlung des Lichtes betreffenden heuristischen Gesichtspunkt. *Annalen der Physik*, 322, 132–148.
Gertner, J. 2013. *The Idea Factory: Bell Labs and the Great Age of American Innovation*, Penguin.
Kirkpatrick, W. E. & Sears, R. W. 1952. *Semiconductor signal translating device*.
Millikan, R. A. 1913. On the elementary electrical charge and the Avogadro constant. *Physical Review*, 2, 109.
Nemet, G. F. & Husmann, D. 2012. PV learning curves and cost dynamics. In: Willeke, G. & Weber, E. (eds.) *Advances in Photovoltaics Part I*, 1st Edition. Academic Press.
NRC 1972. *Solar Cells: Outlook for Improved Efficiency*, National Academies, National Research Council (NRC) Ad Hoc Panel on Solar Cell Efficiency.
NYTimes. 1954. Vast power of the sun is tapped by battery using sand ingredient. *New York Times*, p. 1.

Ohl, R. S. 1946. *Light-sensitive electric device.*
Ohl, R. S. 1948. *Light-sensitive electric device including silicon.*
Paley Commission 1952. *The Promise of Technology—The Possibilities of Solar Energy.* Washington, DC, President's Materials Policy Commission.
Palz, W. 2010. *Power for the World: The Emergence of Electricity from the Sun*, Pan Stanford Publishing.
Perlin, J. 1999. *From Space to Earth: The Story of Solar Electricity*, Ann Arbor, MI, AATEC Publications.
Prince, M. 1955. Silicon solar energy converters. *Journal of Applied Physics*, 26, 534–540.
Scaff, J. H. 1946. *Preparation of silicon materials.*
Sparks, M. 1953. *Method of making p–n junctions in semi-conductor materials.*
Teal, G. K. 1947. *Photoelectromotive force cell of the silicon-silicon oxide type and method of making the same.*

4
US TECHNOLOGY-PUSH

The 1970s transformed PV technology and the solar industry by making energy issues a top social priority for nearly a decade. Cell efficiencies tripled, costs fell by a factor of five, innovative policies were adopted, and thousands of people entered the field. This burst of activity was concentrated in the US, which continued its role as the global leader in PV from the 1940s until the mid-1980s when the US federal government purposively dropped PV as a priority technology and leadership shifted to other countries.

The Arab Oil Embargo in October 1973 quadrupled the price of oil over three months, putting energy issues instantly at the center of public debates in the US, Japan, and other countries affected by the price shock. President Nixon launched Project Independence and made energy the centerpiece of his 1974 State of the Union address. This change was dramatic in that energy had never existed as a policy area before. Although specific challenges with particular fuels at certain times got attention, there was nothing close to Nixon's comprehensive approach.

Although PV was never more than on the sidelines of the US government response to the pressing energy security challenges, the crisis was still transformative for the industry. The embargo opened a massive policy window, which the industry was well poised to take advantage of due to efforts that preceded the crisis. Momentum had been building in solar before the embargo and the Cherry Hill Conference on R&D priorities for PV took place fortuitously a week after the embargo. The US Federal government primarily adopted a "technology-push" approach, investing $1.7 billion in PV R&D from 1974 until 1981 when budgets were slashed. Ninety percent of that $1.7 billion was under the four Carter Administration budget requests. While much progress was made, ambitious plans to make PV competitive with oil-fired electricity production by 1986 (Maycock, 1981) and grid electricity by 1990 (Roessner, 1984) were far from met.

Key US institutions were established during this period including the Department of Energy (DOE) and the Solar Energy Research Institute (SERI), which was later renamed the National Renewable Energy Laboratory (NREL). In addition, novel policies such as the public procurement program known as the Block Buy and an early version of a feed-in tariff in California called the Interim Standard Offer Contract #4 (ISO4) emerged during this time. The combination of federal funding, new institutions, procurement, and expectations that PV's time for commercialization had arrived, entrained thousands of people into the field and stimulated the industry to become more international, more cost conscious, and more specialized.

The US program attracted people from around the world to work in the US, to contribute to the US effort, to visit, and to study. The federal emphasis on solar peaked in 1979 with the highest PV R&D budget in 1980. Thereafter, PV became a lower priority and with the election of President Reagan in 1980, the US PV program was systematically dismantled. Combined with the crash in energy prices in the mid-1980s, the remaining firms in the industry were left to survive on niche market applications. As a result, the knowledge developed in this period—on the technology, on manufacturing, and on policy design—migrated to countries like Australia, Germany, and Japan, where prospects for government support and for markets looked far brighter.

From a National Innovation Systems (NIS) perspective, the US followed its highly successful strategy that began after the Second World War with the notion that public investment in scientific research is the key to developing the technologies that improve society and bolster economic growth (Bush, 1945). This strategy was used to justify the establishment of the National Science Foundation (NSF) (Bush, 1947). This idea, sometimes known as "technology-push" (Rothwell, 1992) was central to the success of the Manhattan Project and the Apollo Project. Indeed, Nixon cited those two programs specifically when he announced Project Independence. The ten billion dollars spent on energy R&D in the Project Independence era was of the same order of magnitude as those other two projects, higher than Manhattan and lower than Apollo (Nemet and Kammen, 2007). A critical review of large US R&D programs pointed to PV R&D in this era as the most successful program of the several cases it analyzed (Cohen and Noll, 1991). The emphasis on high technology and military applications was strongly representative of the US NIS. PV in this period, especially in the late 1970s, also revealed a split in conceptions of where the US NIS should be headed. The demand-pull advocates pursued an activist strategy in which the government should become active in procuring early technology, as had been successful in jet engines and semi-conductors (Nelson and Langlois, 1983; Ruttan, 2006; Mowery, 2009). A distinct approach emerged with the Reagan Revolution of 1980, which advocated a return to the Vannevar Bush technology-push emphasis. However, the Reagan approach was a considerably limited version of technology-push, in which the government should fund only basic research and leave it to the market to create opportunities for those nascent technologies. The Block Buy program was an initial demand-pull program that in 1981 was cut severely to move to a

Reagan-era technology-push-light strategy, in which the government focused only on basic research.

The Arab Oil Embargo and Project Independence

On October 6, 1973, Egyptian and Syrian forces launched air strikes and artillery against Israeli positions in the Sinai Peninsula, followed by an invasion of 32,000 infantry. Their goal, in what became the Yom Kippur War was to recapture the Sinai, which Egypt had lost to Israel in the 1967 Six-Day War. A week later, President Nixon authorized Operation Nickel Grass, in which the US Air Force flew in aircraft, tanks, artillery, and ammunition to Ben Gurion airport near Tel Aviv. In response, the Organization of Arab Petroleum Exporting Countries (Arab OPEC members plus Egypt and Syria) announced an embargo—they would stop selling oil to the US, the UK, the Netherlands, Canada, and Japan. OAPEC countries also committed to reducing production overall, which they accomplished by the end of that year with a 25% reduction. Following this announcement, the price of oil rose by a factor of four between October and January.

The events in the Middle East in fall 1973 quickly got the world's attention and the connections to energy policy were immediate—President Nixon gave the first speech on Project Independence just two weeks after the Embargo began in an Oval Office broadcast (Nixon, 1973). However, a series of efforts that preceded the embargo made way for the possibility of a swift reaction as soon as the crisis opened a policy window (Figure 4.1).

It was not an entirely new focus. For example, some emerging PV activity was already occurring, like Crystal Systems which started in 1971 to produce large silicon crystals. In addition, solar was receiving positive press in 1971–72 (Laird, 2001). On energy more generally, the Paley Commission had warned about relying on Middle Eastern oil in 1952. Nixon presented a plan to Congress for an energy agency in June 1971. He had lifted the oil import ban in 1970, leading to a dramatic increase in imports thereafter. Indeed, just prior to the crisis, the Dixy Lee Ray report to Nixon had warned that energy consumption would double in the next decade and that importing substantial quantities of oil would ensue (Ray, 1973).

The most important precursor event, however, was the Cherry Hill Conference, which was held in Cherry Hill, NJ from October 23 to 25, just two weeks after the Yom Kippur War began. In an interview industry veteran Charlie Gay calls it,

"the most catalyzing event on the development of the PV industry in the US."

The Cherry Hill Conference was a joint NSF-NASA conference to develop goals for PV R&D across a wide range of activities, including new materials, testing equipment, and large-scale systems. It was part of the NSF Research Applied to National Needs (RANN) program that later established the Flat Plate Array Solar Array Project (FSA). The conference consisted of six working groups of 130 attendees with broad representation: 45% industry, 35% academic, 20% government

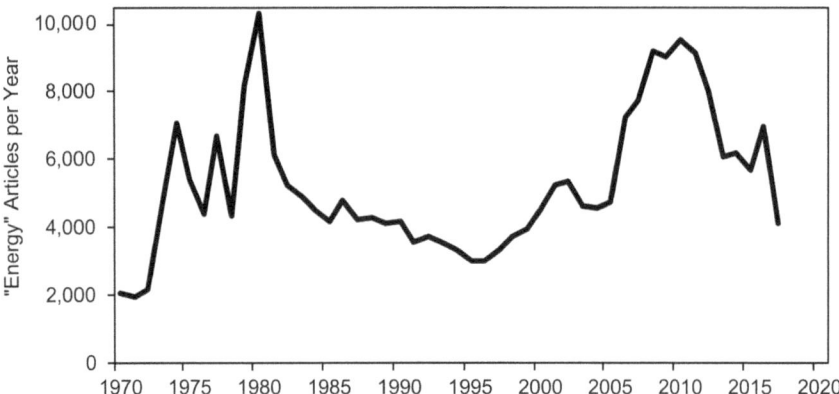

FIGURE 4.1 Trends in the salience of energy as a social concern. Counts of the word "energy" in articles appearing in the *New York Times* 1970–2017.

(Blieden, 1999), mostly of people with experience working on PV for space applications. Aware of the geo-political context in which they were meeting, those at the conference nimbly assembled an R&D budget for a crash program in PV, which also included milestones, goals, and cost targets. They proposed $250 million in R&D from 1975 to 1985. That budget was later included in the Project Independence blueprint of 1974, although once implemented the 1975–85 total was four times what seemed like an already ambitious proposal at Cherry Hill.

Nixon's November 7 speech followed just two weeks after the Cherry Hill Conference,

> Let us set as our national goal, in the spirit of Apollo, with the determination of the Manhattan Project, that by the end of this decade we will have developed the potential to meet our own energy needs without depending on any foreign energy sources. (Nixon, 1973)

The Arab Oil Embargo opened a gaping policy window. It created "energy" as a new policy area, one that had barely been considered a national concern, other than the acute awareness in the military about access to fuel during conflicts. There was no Department of Energy and no one had talked about energy in a policy context. Some energy data were collected, and the military knew the importance of access to fuel, but most of the information regarding it was hidden in the private sector. A big change throughout the industry was making energy a public issue from which energy information became a public good. President Nixon made Project Independence a centerpiece of his State of the Union Address on January 30, 1974.

> In all of the 186 State of the Union messages delivered from this place, in our history this is the first in which the one priority, the first priority, is energy …

As we move toward the celebration two years from now of the 200th anniversary of this Nation's independence, let us press vigorously on toward the goal I announced last November for Project Independence. Let this be our national goal: At the end of this decade, in the year 1980, the United States will not be dependent on any other country for the energy we need to provide our jobs, to heat our homes, and to keep our transportation moving. To indicate the size of the Government commitment, to spur energy research and development, we plan to spend $10 billion in Federal funds over the next 5 years. That is an enormous amount. But during the same 5 years, private enterprise will be investing as much as $200 billion—and in 10 years, $500 billion—to develop the new resources, the new technology, the new capacity America will require for its energy needs in the 1980's. That is just a measure of the magnitude of the project we are undertaking. (Nixon, 1974)

PV research was to become part of the $10 billion, even if a minor part compared to coal, nuclear power, and synthetic fuels. Despite the urgency of the moment, Congress did not move quickly on ERDA due to the growing Watergate scandal that led to Nixon's resignation that August. Instead, Nixon moved on his own. In December 1973 he created the Federal Energy Office and by April the leaders had hired 1,500 employees in Washington and 1,000 in ten regional field offices. In June that organization became the Federal Energy Administration with the mandate to come up with a long-term energy plan, which was released five months later with Gerald Ford as President.

The "Project Independence Blueprint" (Kelly et al., 1974) had a few pages specifically about solar energy. Rather than detailing the federal role, it focused on justifying the large role for private sector investment that Nixon had promised. It forecasted one trillion dollars in energy sector investments from 1974 to 1990 and that 0.8% of that would be for solar—and that small amount was mainly for solar heating not photovoltaics. Even though the blueprint was a long-term plan with detail out to 1990, it was dismissive of PV:

For the foreseeable future, solar cells will be used in special applications where they do not have to compete directly with utility-supplied power. These applications include power on remote signal equipment (e.g., ocean buoys), for installations now being served by small engine-driven generator sets, and for small-scale charging of batteries on boats, at summer camps, etc. (Kelly et al., 1974)

PV was not part of the response to the US's energy import crisis, even in the long term. Nuclear had a much larger role with plans for 1,000 new reactors by 2000—although less than thirty were built after the report was released.

Political Scientist Frank Laird argues that the crisis provided an opportunity to revise the existing problem frame and institutionalize a new set of ideas to guide that frame (Laird, 2001). But core ideas changed very little and the new institutions were never realized. PV did not fit into the problem frame, which focused on large, well-established technologies like coal and nuclear. Despite an emerging advocacy coalition for PV consisting of environmentalists and technologists, PV never achieved "technological citizenship" in the 1970s (Laird, 2001).

New institutions: ERDA, SERI, and DOE

However, institutions did emerge that had a strong effect on the long-term development of PV, i.e. beyond the Blueprint's definition of 1990 as long-term. First, with Nixon's impeachment behind them, Congress created the Energy Research and Development Administration (ERDA) in October 1974 as part of the Energy Reorganization Act of 1974. Robert C. Seamans served as the first head of ERDA with Bob Fri as deputy. ERDA quickly released publications such as, "Creating Energy Choices for the Future," National Energy RD&D Plans in June 1975, and the "National Solar Energy RD&D report" (Reuyl, 1977). Paul Maycock left his position in Strategic Planning at Texas Instruments and arrived at ERDA in June 1975 to run the PV program. But the role of PV at ERDA remained minor until the Carter Administration began in 1977.

The second important institution was created when ERDA authorized the Solar Energy Research Institute (SERI) in 1977. Carter succeeded in encouraging Congress to elevate energy to a cabinet position, something that both Nixon and Ford had attempted without success (Laird, 2001). The Federal Energy Agency became the Department of Energy in the first year of the Carter Administration. James R. Schlesinger was sworn in as first Secretary of Energy in October of that year. Carter also elevated solar within energy by moving much of the ERDA solar program to SERI. Formally, the solar program stayed in the DC offices of the DOE, which also gave out money to universities and industries for PV R&D (Laird, 2001). Paul Rappaport was the program's first director of a staff of 75, located in Golden, CO. The Solar PV R&D Act of 1978 further raised the prominence of these nascent institutions. The Act established the "National PV Program," a 10-year, $1.5 billion plan for PV R&D with a focus on near-term commercialization and a goal of 50 GW by 2000—a goal attained in 2011. It authorized three directing organizations (by technology): NASA Jet Propulsion Lab (JPL) for crystallized silicon arrays, SERI for thin films and high efficiency silicon, and Sandia National Laboratory to work on concentrating PV and systems. As Maycock recalled in an interview,

> "The PV plan of 1978 called for technology readiness in 1982 and commercial readiness in 1986—leading to shipping competitively priced product in 1988. In other words, we were 10 years from economic PV for central power."

The Iranian Revolution in 1979, and the ensuing second oil crisis, led Carter to push solar even further. On June 20, 1979 President Carter announced a program to increase the nation's use of solar energy, including a solar development bank and increased funds for solar energy R&D. The White House coordinated a large interagency review and recommended an aggressive goal to implement and use current solar technologies. The publication of the "Domestic Policy Review of Solar Energy" (United States Department of Energy, 1979) was perhaps the highest PV has ever gotten as a priority of the US federal government. However, the second half of the Carter administration placed much less emphasis on solar. Carter realized that solar, and renewables generally, could not do anything for him in the short term and he began to propose cuts to solar R&D spending at the end of his term. Carter's "Crisis of Confidence" speech on July 15, 1979 claimed that:

> Energy will be the immediate test of our ability to unite this nation, and it can also be the standard around which we rally. On the battlefield of energy we can win for our nation a new confidence, and we can seize control again of our common destiny. (Carter, 1979)

But solar had only a minor role in that vision, even though he briefly restated his target for solar (by which he meant renewables) as 20% of total energy generated by 2000 through the never-funded solar bank. Still, much was accomplished during the burst of PV work that began in 1975. The cornerstone of which was the Flat Plate Solar Array Project.

Block buys

In 1975, ERDA launched the world's first demand-pull program for PV. ERDA assigned management of the program to the Jet Propulsion Lab (JPL) of Pasadena, CA, which had a strong semi-conductor group and experience in developing PV for space applications (Blieden, 1999). ERDA called the program the "Low-Cost Solar Array Project," which eventually was renamed as the "The Flat-Plate Solar Array Project" (FSA). Part of that program included the "Block Buy" Program. The Apollo Project thinking that pervaded Project Independence was prominent in the Block Buy Program. At the time, exuberance among leaders of the program planned for competitive PV by the mid-1980s, not 2000 as earlier estimates predicted (Maycock, 1981). Just as it was ramping up in the early 1980s, the Reagan Administration cut the PV R&D budget from $130 million per year to $50 million per year and made clear that the government would focus on basic research rather than "applied research"—nothing could be more applied and still be called research than the Block Buy program. The FSA was ultimately phased out by the end of September 1986.

The motivation for the Block Buy Program was directly linked to Project Independence—to develop module and array production technology and know-how that would be needed to create the widespread terrestrial use of photovoltaics by 1985. In 1974 PV technology had been around for 20 years, but was still too expensive and limited to serving about 100 kW per year of niche markets, such as off-grid houses, communications, marine navigation aids, and satellites. ERDA wanted to develop a real industry, one that eventually would allow PV to compete with established fossil fuel power generation, particularly oil, which accounted for a substantial amount of

72 Creating a technology

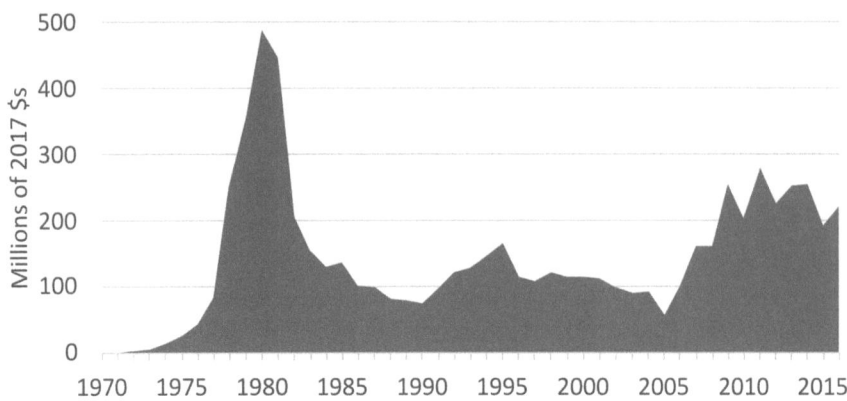

FIGURE 4.2 US Federal PV R&D Budgets (millions of 2017$s).

electricity generation and had recently become scarce. JPL was intentionally chosen to develop terrestrial PV using their know-how from space applications.

To achieve widespread adoption of PV, the project planned to develop solar module technology with 10% efficiency, a 20-year lifetime, and a selling price of $0.50 per Watt (in 1975 dollars). Cost-cutting was pursued at all phases of production, ranging from silicon purification to verification of cell reliability and performance. When a new technical option arose from R&D, economic analyses were performed and only the most promising were continued. The majority of FSA funds were used to sponsor R&D in private organizations and at universities, which led to an effective federal government-university-industry team that facilitated the rapid advance of PV technology.

The cost-cutting goals were closely related to the experience curve concept championed by the head of the ERDA PV program, Paul Maycock. Maycock had used the experience curve concept extensively in his role in strategic planning at Texas Instruments (TI) from 1964 until he moved to the solar program in 1975. TI used the experience curve to project costs for diodes, transistors, integrated circuits, and later for calculators. The experience curve pulled together the technical, engineering, and cost considerations on its y-axis while also including marketing and price elasticity on its horizontal axis. Both engineers and marketing departments found it useful for product strategy. Maycock brought the experience curve to ERDA and, soon after, developed a projection of PV costs (Maycock and Wakefield, 1975) that is remarkably on-target, even over forty years later (Figure 4.3). Included in that prediction is indication of a subsidy program of $10b–20 billion to move PV to 100 GW scale. Put in 2017 dollars, that investment would be about $50 billion, or about a quarter of what the Germans spent to accomplish that goal (Chapter 6). On reflection, Maycock says that the experience curve was never fully embraced by ERDA and later DOE management. The government did not exert the same control over its production and marketing decisions as a single firm like TI did. It also was considered "planning" which sounded suspiciously like communism to some and after the 1980 election, was not tolerated. The related shift away from

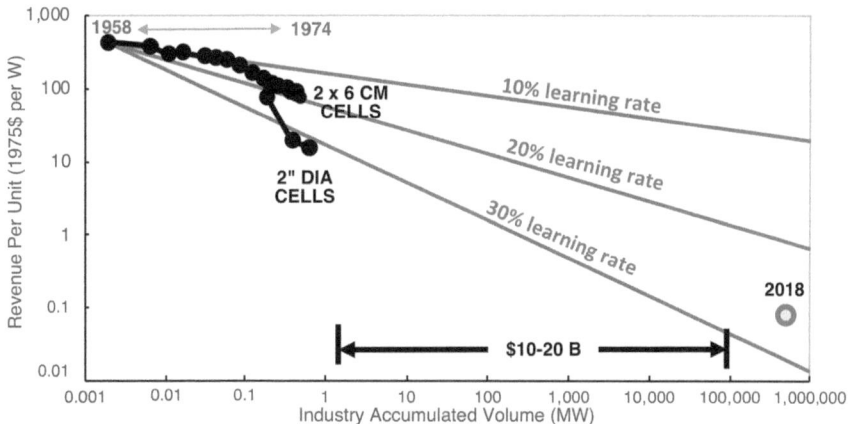

FIGURE 4.3 Learning curve forecast for PV modules from 1975.

applied research to basic research further distanced policymakers from a planning tool. Costs and deployment goals became guiding principles rather than targets and technical parameters like cell efficiency became more prominent objectives. Nonetheless, those working on the FSA program developed very detailed cost goals, aided in part by a cost modeling computer program called the Solar Array Manufacturing Industry Costing Standards (SAMICS).

The Block Buy was a public procurement program, in which ERDA agreed to purchase pre-specified amounts of PV from private firms. One of the central questions the designers of the program faced was how much to buy. They arrived at their planned number—about 100–300 kW per year—by telling manufacturers their budget, deciding on a price objective of $5 per Watt, and then informally pre-negotiating contracts, leaving the formal procurement process for later. Over the ten years of the program, ERDA bought a little over 400 kW of PV modules from a dozen or so manufacturers. The program accounted for 17% of global PV purchasers from 1976 to 1981.

The buys were staged in a series of five "blocks," with increasingly demanding specifications for the manufacturers (see Table 4.1) (Roessner, 1982; Maycock and Stirewalt, 1985; Green, 2005; Wolfe, 2018). Cost-cutting was enabled by SAMICS, a bottom-up computer model which looked at every step in the production process and targeted opportunities to reduce costs. Although the plan was to steadily increase the buys, the budget cuts that were implemented in 1981 drastically cut the program in Buys 4 and 5.

The Block Buy Program entrained a large swath of the PV industry in its efforts:

> at one time or another … almost everyone in the [PV] community was a working partner in the attempt to achieve [the FSA Project's] goals. (Callaghan and McDonald, 1986)

TABLE 4.1 US Block Buy Program

Block	Years	Purchases (kW) Planned	Actual	Budget $ millions	$/W	Producers
1	1975–76	46	54/58	0.9	15.22	M7 International, Sensor Technology, Solarex, Solar Power Corporation, Spectrolab
2	1976–77	90	123/127	1.7	13.72	Sensor Technology, Solarex, Solar Power Corporation, Spectrolab
3	1978–79	200	205/259	2.2	10.50	Arco Solar, Motorola, Sensor Technology, Solarex, Solar Power Corporation
4	1980–81	500	26/32			Arco Solar, Applied Solar Energy Corporation, General Electric, Motorola, Photowatt, Solarex, Spire Corporation
5	1981–85	1000	9			Applied Solar Energy Corporation, Arco Solar, General Electric, Mobil Solar, Motorola, Solarex,
Total	1975–85	1836	427			

Sources: Roessner, 1982; Maycock and Stirewalt, 1985; Maycock and Stirewalt, 2005; and Wolfe, 2018

Corporate, academic, and individual interest in the program was widespread. Dr. Morton Prince, once the head of DOE's PV department and former member of Bell Labs' Solar Battery Project, and National Photovoltaics Program directors Paul Maycock and Robert Annan served strategic roles in the program's success.

The plans for competitive PV by 1986 shifted in 1981, not only due to the November 1980 election of Ronald Reagan, but also because of a less-favorable economic environment for photovoltaics due to low energy prices, which had resulted from energy conservation programs, a shift away from oil for power generation, and new energy supplies:

> Although Congress wished to emphasize demonstration projects, purchases and other market development activities, DOE plans continued to emphasize research and development. The DOE's National Photovoltaic Program Plan, dated February 3, 1978, envisaged as one alternative an 8 year, US $380 million procurement program (a "market-pull" initiative), but this program never went beyond the planning stages. (Roessner, 1982)

In response, the Block Buy program set its sights on long-range efforts, which the private sector avoided due to increased risk and longer payoff times. To be competitive against established utility generation plants in the 1990s, it was estimated that PV-generated electricity would have to be sold at $0.17 per kWh (in 1985 dollars). Efficiency targets were raised from the initial 10% goal to between 13 and 17%. Necessary module life expectancy was extended to 30 years.

The summary report on the project claims that Congressional leaders described the FSA as

> one of the best managed programs every undertaken by the government (Christensen, 1985)

Indeed, the accomplishments of the program are many. As the FSA was winding down, the project analysis and integration team evaluated the leading PV technologies under development at the time and found that the FSA project had achieved many of the goals set by the program and had made significant strides toward the redirection that took place in 1981. While PV ultimately did not become widespread in the US by 1985, the FSA should be considered an overall success in terms of its positive influence on the trajectory of the solar energy industry. The program succeeded in addressing critical bottlenecks in the manufacturing process. It was not so much about reducing costs as it was demonstrating the reliability of the technology, as well as ensuring the private sector was making investments in manufacturing and had real stakes in the outcomes. It took the government creating a market for those investments to happen. Companies competed to see which would be the first to produce 1 MW per year. Technically, the cells that emerged in Block Buy V are quite similar to those of 2018. Module efficiency improved from 5% to 15%, and by 1985, prices were down to 15% of the 1974 level. Ten-year product warranties were becoming available. It was also important that the program led to a diversity of companies and enabled specialization in the value chain. For example, several companies emerged to produce poly-silicon such that it became a commodity. Moreover, none of the module manufacturers were producing purified silicon in house—similarly module producers were not slicing wafers either because specialty suppliers had emerged.

The tide turns

Two events reversed the momentum that had been building in the US PV industry between Cherry Hill in October 1973 and the 1979 Domestic Policy Review. First, Ronald Reagan's election in November 1980 led to a conservative movement that sought to shrink non-military government activities and PV seemed to be especially picked out for cuts. Reagan had made clear his view on federal efforts to deal with energy problems during a debate with Carter in October:

> I just happen to believe that free enterprise can do a better job of producing the things that people need than government can. The Department of Energy has a multi-billion-dollar budget in excess of $10 billion. It hasn't produced a quart of oil or a lump of coal, or anything else in the line of energy. (Reagan and Carter, 1980)

Of course, DOE's mandate was mainly to support research and innovation and it was not in the business of producing energy. In fact, most of the agency's budget goes to nuclear weapons and basic physics research, rather than R&D. The conservative movement that Reagan led rejected government planning. They also rejected conservation and, for example, repealed the 55-mph speed limit that Nixon had enacted. The emphasis was on increasing production and expansion of US presence in the Persian Gulf, which had been started by Carter, who spelled his intentions to defend the Persian Gulf with the Carter Doctrine, a response to the Soviet's invasion of Afghanistan. Industry experts spoke of a "scorched earth" vindictive campaign against solar. Despite the early uses in the military and the independence enabled by off-grid systems, PV seemed to fly in the face of conservative thinking. Solar advocates were characterized as a special interest and left-wing ideologues. The Reagan administration said the DOE should be funded at lower levels, stick to basic research, and end commercialization efforts like the Block Buy Program.

> "In 1980, DOE started four 3-year contracts on manufacturing methods for PV cells. The contracts were all terminated in 1981, after President Reagan took office" – Paul Maycock.

Reagan appointed a dentist, James B. Edwards, as Secretary of Energy with orders to shut down the DOE. He fired Dennis Hayes of SERI and by 1985 cut its budget to one fifth of its 1981 level. Second, in October 1985 Saudi Arabia flooded the oil market with excess production causing the price of oil to collapse and subsequently stay low, a 20-year body blow for alternative energy. Oil independence was no longer an issue, Americans bought fuel-intensive sport utility vehicles and imports soared. In 1986 Reagan had the White House solar hot water panels removed, which Carter had installed in 1979.

California, PURPA, and Standard Offer Contracts

While the US federal government in the 1970s made clear that the focus of PV activity worldwide was at ERDA and SERI, concurrent developments in California also made major contributions to the development of PV. First, California passed Assembly Bill 1575, the Warren-Alquist Act in 1974 (Taylor et al., 2006), which established statutory authority to the California Energy Commission (CEC). The CEC would play a central role in funding research and deployment of PV for over 40 years. Second, the Carter Administration passed the Public Utilities

Regulatory Policy Act (PURPA) in 1978 (Hirsch, 1999). PURPA required electric utilities to purchase power from small producers, "qualifying facilities," at a price equal to the utilities "avoided cost" of producing that electricity. It was then left to states to implement these purchase agreements. The California Public Utility Commission (CPUC) led the way by developing a series of standard contracts to facilitate these agreements. The utilities proposed prices that were set by the price of oil, which they were still using to fire peak generation. Standard Offer Contracts 1 and 2 were tied to oil price and would, thus, fluctuate dramatically. Wind power developers, the most active qualifying facilities at the time, argued that wind turbines and other renewable energy projects were long-term capital investments with low marginal costs. They argued that it would be better to have a fixed price so that they could guarantee a return to capital and earn a profit. The CPUC agreed and developed Interim Standard Offer Contract #4 (ISO4) in 1985 with a fixed payment schedule. Because expectations in the first half of the 1980s were that energy prices would continue to rise, these contracts included annual price rises of 7% per year. The ISO4 led to the California wind power boom, in which Wall Street Banks financed a billion dollars of wind power projects (Nemet, 2009). A similar amount of solar thermal electric (STE) plant capacity was also built under ISO4s (Lotker, 1991). In the mid-1980s, PV was still too expensive to benefit from the generous ISO4 contracts. Many of the wind and STE developers went bankrupt (Shinnar, 1993) as the "interim" aspect was invoked and the contracts renegotiated at much lower levels once oil prices crashed in 1985.

Still, people around the world were watching. It became clear to all that fixed price contracting could mobilize very large amounts of capital quickly—even for novel technologies like wind and STE that were competing with large established technologies run by monopolists in the electric utilities. It is difficult to pinpoint a German delegation visiting the CPUC or the CEC in the mid-1980s. But Germans interviewed in this project made clear that they were closely observing developments in California's ISO4s and they saw that fixed price contracting worked. The Danes who were supplying the bulk of the wind turbine equipment that generated the power for these contracts were also well aware of the policy design that effectively created a market. Seven years later, Aachen, Germany began to propose its own ISO4, in that case for PV (see Chapter 6).

Outcomes

Even though they were cut just as they seemed to be building momentum, the ERDA and SERI programs produced a variety of outcomes. They advanced the technology, drew people into the field, and developed an international industry.

Some of the technical improvements advanced the knowledge frontier and raised the state of the art. The programs developed a broad set of technological paths and demonstrated an array of feasible products (Maycock, 1981). One study identifying the twenty most important breakthroughs from 1954 to 2004 (Green, 2005) found that 14 were in the 1974–81 era, almost all of which were in the US.

Another study of breakthroughs from 1957 to 1993 identified 39 patents covering 23 breakthroughs in PV technology (Nemet and Husmann, 2012). More than half were developed in the ERDA/SERI period. Examples include: pulse annealing, aluminum-based pastes, hydrogen plasma passivation, and quasi-square wafers. There were advances outside the US at this time too. Martin Green's group at the University of New South Wales developed fine contacts, antireflection, surface texturing, rear contacts, and oxide passivation. Swiss firm HCT Shaping Systems developed wire saws to slice wafers that used wires developed for radial tires instead of piano wire. Impregnated with diamond dust, these saws produced thinner wafers with less loss of material between wafers.

Other improvements were of a more disruptive innovation approach. For example, Siemens invented the multi-crystalline silicon ingot process in the late 1970s. Previously all PV was grown from large single crystals. The multi-crystalline technology was developed and commercialized in the 1980s and is now prevalent. More broadly, that example made clear that solar does not need the high purity that computer semi-conductors need, which can only use single-crystalline silicon. Solar could live with lower purity because they could make wafers from poly-crystalline. These innovations set the stage for other examples of disruptive innovation in PV (Wilson, 2018).

Cost reductions were a prominent goal and a signature accomplishment of the program. Costs of modules dropped dramatically, by a factor of five between 1974 and 1981. That drop was the second fastest pace of cost reductions over a 7-year period in the history of PV, exceeded only by a slightly faster fall in the post-2010 period. On the other hand, the program goals were for modules of $0.70 per Watt by 1986 and $0.15–0.40 per Watt by 1990 (Maycock, 1981). The pace of cost reductions slowed in the 1980s and those goals remained far out of reach. Module costs in 1990 were ten times higher than the upper-end target. Modules priced at $0.40 per Watt were not available until 2012. Still, the cost outcomes are impressive in that the cost reductions were not a result of massive scale that characterized post-2010, but rather essentially an R&D program combined with a relatively modest public procurement effort. Perhaps what drove much of this was the extensive private sector involvement that arose, not from trying to take advantage of the availability of public R&D funds, but rather as part of a broader consensus that solar's time to commercialize had come.

Another outcome was that the federal emphasis—as well as the broader public focus on energy—drew people in to work on solar. By 1981, 4,000 people were working in PV (Maycock, 1981). Martin Green of the University of New South Wales toured the US with his lab in 1976 and visited a dozen solar manufacturers to see their labs and understand how they were using equipment like furnaces. Green would go on to set world records in cell efficiency and train a substantial cadre of those who went on to found solar giants in China. Markus Real visited the Rancho Seco PV installation in 1986—he went on to advocate for the world's first rooftop program, Project Megawatt, in Switzerland. These solar advocates came from many directions and the vast majority never left the industry. As former

BP Solar CEO Phillip Wolfe says, "they got hooked," so that even when the industry had its recurring downturns, knowledge was retained in the heads of those who stayed, and most did (Wolfe, 2018). Why did they stay? For one, the technology was fascinating. They also felt like they were working on something good for society—first to address energy security and later for environmental reasons. Many in the industry had strong intrinsic motivation for working on solar that kept them at it, despite the volatile ups and downs of the industry.

A new industry

The burgeoning industry was one of the key outcomes of the federal emphasis on PV in the second half of the 1970s, particularly from 1977 to 1981. A set of viable solar companies emerged and sales by US PV firms went from $7 million in 1976 to $100 million in 1982 (Roessner, 1984). These companies made technology improvements on their own, and, helped by the Block Buy, began to set up manufacturing lines with a more streamlined process. Producers in early 1980s were not firms who met space industry demand in 1960s–1970s but rather new firms who focused on lower costs instead of high performance.

For example, Solarex in Maryland began operations in the summer before the Arab Oil Embargo, increased their efficiency from 10% to 14%, and were early adopters of poly-crystalline silicon (Perlin, 1999). Amoco bought them in 1983. Bill Yerkes founded Solar Technology International (STI) in 1975 to sell cells to the space program. Arco bought STI in 1977 and ArcoSolar became the largest producer in the world from 1983 through the early 1990s when Siemens bought them. In 2002, Siemens sold them to Shell who then, in 2006, sold them to Solarworld. Among other innovations, they commercialized screen printing for metal contacts, gluing cells to glass instead of coating them in silicone. While working at Exxon in the late 1960s, Elliot Berman flew to Japan to buy cells from Sharp and used them to power navigational lighting on oil platforms for the oil company. He later founded Solar Power Corporation, which dropped production costs dramatically by avoiding some of the expensive steps used for space applications such as polishing surfaces.

Until the oil price crashed in 1985, solar companies were able to raise substantial capital in equity markets: Spire Corp listed shares in 1983 and Energy Conversion Devices listed on the NASDAQ in 1985. Right on the cusp of a rapidly declining concern with energy that followed the 1985 price crash, Stanford professor Dick Swanson founded SunPower corporation in 1985. SunPower corporation drew on the R&D efforts of the 1970s and made the most efficient commercial modules in the world for many years. The timing was not opportune and they were unable to raise funds for the new company for six years. Instead they existed on niche market applications, even further afield from bulk electricity than US space applications and later Japanese consumer electronics. They used their light-sensitive semi-conductor materials for infra-red data ports, an early method to wirelessly transmit data between laptop computers. They got their breakthrough in another niche application, an order from Honda for a solar race car. Solar PV survived over and over again, leaving behind tacit knowledge of the industry. This could be done because the technology was modular and flexible enough to serve an array of niche markets when policy, politics, and markets were not in their favor.

Industry response to the Reagan-era changes was mixed. Some saw it as a way for "the industry to stand on its own." While others feared it would lead to increased competition from European and Japanese firms, who were being subsidized.

> After the dramatic improvements in module design over a 10-year period achieved as a result of the well-orientated US Government program outlined above, declining funding for photovoltaics in the USA shifted the key developments elsewhere. (Green, 2005)

Indeed, innovation in PV also shifted elsewhere, especially to the emerging super power, Japan. Markus Real began Project Megawatt in Switzerland in 1987. He visited the US a couple of years earlier, had seen solar projects in California, and had a girlfriend who worked at SERI. German conglomerate Siemens bought Arco Solar for $40m in 1989, which was also the first year that German PV R&D exceeded that of the US.

References

Blieden, R. 1999. Cherry hill revisited—a retrospective on the creation of a national plan for photovoltaic conversion of solar energy for terrestrial applications. *AIP Conference Proceedings, 1999*. AIP, 796–799.

Bush, V. 1945. *Science The Endless Frontier: A Report to the President by Vannevar Bush, Director of the Office of Scientific Research and Development*. Washington, DC, United States Government Printing Office.

Bush, V. 1947. A National-Science-Foundation. *Science*, 105, 302–305.

Callaghan, W. & McDonald, R. 1986. *The Flat Plate Solar Array Program Final Report*. Jet Propulsion Laboratory.

Carter, J. 1979. *The "Crisis of Confidence" Speech*. The White House.

Christensen, E. 1985. *Electricity from photovoltaic solar cells: Flat-Plate Solar Array Project of the U.S. Department of Energy's National Photovoltaics Program, 10 years of progress. JPL publication: 400–279*, Pasadena, CA, Prepared by the Jet Propulsion Laboratory for the U.S. Dept. of Energy through an agreement with the National Aeronautics and Space Administration.

Cohen, L. R. & Noll, R. G. 1991. *The Technology Pork Barrel*, Washington, Brookings.

Green, M. A. 2005. Silicon photovoltaic modules: A brief history of the first 50 years. *Progress in Photovoltaics: Research and Applications*, 13, 447–455.

Hirsch, R. F. 1999. *Power Loss: The Origins of Deregulation and Restructuring in the American Electric Utility System*, Cambridge, MA, The MIT Press.

Kelly, J. T., White, D. L. & United, S. 1974. *Project Independence: Federal Energy Administration, Project Independence blueprint: Final Task Force Report*. Washington, DC, For sale by the Supt. of Docs., U.S. G.P.O.

Laird, F. N. 2001. *Solar Energy, Technology Policy, and Institutional Values*, New York, Cambridge University Press.

Lotker, M. 1991. Barriers to Commercialization of Large-Scale Solar Electricity: Lessons Learned from the LUZ Experience. Sandia National Laboratories.

Maycock, P. 1981. Overview of the US Photovoltaic Program by an Ex-DOE Person. Proceedings for 15th IEEE Photovoltaic Specialists Conference. Kissimmee, FL, 1981.

Maycock, P. D. & Wakefield, G. F. 1975. Business Analysis of Solar Photovoltaic Energy Conversion. 11th IEEE Photovoltaic Specialists Conference, May 6–8 1975New York. IEEE, 252–255.

Maycock, P. D. & Stirewalt, E. N. 1985. *A Guide to the Photovoltaic Revolution*, Emmaus, PA, Rodale Press.

Mowery, D. 2009. National security and national innovation systems. *The Journal of Technology Transfer*, 34, 455–473.

Nelson, R. R. & Langlois, R. N. 1983. Industrial Innovation Policy: Lessons from American history. *Science*, 219, 814–818.

Nemet, G. F. 2009. Demand-pull, technology-push, and government-led incentives for non-incremental technical change. *Research Policy*, 38, 700–709.

Nemet, G. F. & Kammen, D. M. 2007. U.S. energy research and development: Declining investment, increasing need, and the feasibility of expansion. *Energy Policy*, 35, 746–755.

Nemet, G. & Husmann, D. 2012. Historical and future cost dynamics of Photovoltaic technology. In: Sayigh, A. (ed.) *Comprehensive Renewable Energy*, Oxford, Elsevier.

Nixon, R. M. 1973. *Speech on Project Independence*. The White House.

Nixon, R. M. 1974. *State of the Union Address*. Washington, D.C.

Perlin, J. 1999. *From Space to Earth: The Story of Solar Electricity*, Ann Arbor, MI, AATEC Publications.

Ray, D. L. 1973. *The nation's energy future. A report to Richard M. Nixon, President of the United States*, Washington, U.S. Atomic Energy Commission.

Reagan, R. & Carter, J. 1980. Carter–Reagan Debate.

Reuyl, J. S. 1977. *Solar Energy in America's Future: A Preliminary Assessment*, ERDA.

Roessner, J. D. 1982. Government–industry relationships in technology commercialization: The case of photovoltaics. *Solar Cells*, 5, 101–134.

Roessner, J. D. 1984. Commercializing solar technology: The government role. *Research Policy*, 13, 235–246.

Rothwell, R. 1992. Successful industrial-innovation – critical factors for the 1990s. *R&D Management*, 22, 221–239.

Ruttan, V. W. 2006. *Is War Necessary for Economic Growth? Military Procurement and Technology Development*, Oxford, Oxford University Press.

Shinnar, R. 1993. The rise and fall of Luz. *Chemtech*, 23, 50–53.

Taylor, M., Thornton, D., Nemet, G. & Colvin, M. 2006. *Government Actions and Innovation in Environ Technol for Power Production: The Cases of Selective Catalytic Reduction and Wind Power in California*. Sacramento, California Energy Commission.

United States Department Of Energy 1979. *Domestic Policy Review of Solar Energy*, U.S. Department of Energy.

Wilson, C. 2018. Disruptive low-carbon innovations. *Energy Research & Social Science*, 37, 216–223.

Wolfe, P. R. 2018. *The Solar Generation: Childhood and Adolescence of Terrestrial Photovoltaics*, Hoboken, NJ, IEEE Press and Wiley.

PART 2
Building a market

5
JAPANESE NICHE MARKETS

In spring 1973, Japan's Agency for Industrial Science and Technology (AIST), within its powerful Ministry for International Trade and Industry (MITI), began reviewing annual proposals for its Large-Scale Projects program. Deputy Director General of Development, Nebashi Masato and research and development head Suzuki Ken noted how many of the proposals involved energy topics[1] (Shimamoto, 2014b). Even though the first oil crisis was still months away, Japanese manufacturers had been urging the government to do something about the country's growing dependence on oil imports. Nabashi, Suzuki, and colleagues graded the energy related themes according to their expected social impact. They gave solar energy an A, the highest possible grade. However, when Nabashi then asked the heads of the various AIST R&D programs who wanted to lead a new initiative on solar energy, not a single person raised their hand. Nabashi asked Suzuki to do so and he too declined saying it would be too difficult. The lack of enthusiasm for leading a grade-A research program was not only caused by excess caution. Rather, there was a well thought through concern that solar energy was many years from becoming commercially viable—and these bureaucrats knew well that the Large-Scale R&D programs at MITI had to show positive results within five years. R&D officials at AIST would need even stronger motivation and a change in performance management timelines to get on board. Both arrived a few months later with the Arab Oil Embargo.

Overview

Japan became a global leader in PV by putting in place all the key innovation components needed to develop a promising technology, commercialize it, create a market for it, and eventually dominate world production of it.

Japanese companies and the national government conducted decades of R&D with a focus on commercial applications from the outset, in the 1950s. Japanese global conglomerates, such as Kyocera, Sanyo, and Sharp, entered the PV industry as early as those in the US and found a sequence of profitable niche markets—particularly consumer electronics—in which to sell PV. As niche markets became saturated, MITI created a rooftop subsidy program with a variety of innovative features: up-front rebates, a declining rebate schedule over ten years, and complementary policies such as net metering. During that program from 1994 to 2005 over 200,000 households installed PV systems and Japanese PV producers scaled up to meet that demand. The leading Japanese firm Sharp increased its production scale by a factor of 200 and by the mid-2000s attained the highest market share any company has had since the early days of the industry. Soon after that, Sharp's growth stalled. German and later Chinese producers were able to produce at much higher volumes. Japan's relinquishing its leadership on solar is surprising given its long history, the level of public commitment, and the deep engagement by firms. But it is not unique; the US and Germany also gave up what seemed like insurmountable leadership positions. Japan lost its lead due to its inability to anticipate the emergence of the German market, its focus on high quality rather than low cost production, a reluctance to enter into long-term contracts for silicon feedstock, the reallocation of production investment to thin film PV, and relinquishing technology development leadership to equipment suppliers.

Japan's national innovation system is distinct in the large role for the state (Johnson, 1982). Government activities were more important for Japan than for any other PV country, even compared to China. The Japanese government not only funded R&D but determined much of the direction of private sector R&D. Japanese R&D rose just as US R&D was severely cut under Reagan. It implemented a strong industrial policy that involved creating shared expectations in the industry and also purposively facilitating inter-technology spillovers. At the center of the Japanese developmental state (Johnson, 1982) was the Ministry of International Trade and Industry (MITI). MITI's distinct approach was to get involved in an emerging technology area from its beginning (Freeman, 1987). This early entry was different from the Chinese central government approach, which is to provide support once companies become internationally competitive. One example is MITI's involvement in reducing transactions' costs, such as encouraging net metering, for installations in the early 1990s before a real market existed.

Japan's national innovation system for PV is also distinct in that it involved large multinational conglomerates with a very minor role for start-up businesses (Shum and Watanabe, 2007). Consequently, Japan's governance of PV was highly corporatist—the government developed policy in consultation with large companies with almost no role for citizen engagement or entrepreneurial input. Contrast this to what we will see in Chapter 6 with Germany, another traditionally corporatist political system where PV emerged from the grassroots and start-up companies became global leaders. This system promoted R&D consortia (Watanabe et al., 2004), which led to shared expectations among participants, for example about future deployment levels and prices. It also involved technology roadmaps, a distinguishing characteristic of Japanese industrial policy and PV development.

Creating shared expectations was an effective way to encourage firms to make large investments in scale.

MITI made intentional efforts to maximize knowledge flows among technology areas. Because the players were large conglomerates, like Sanyo and Kyocera, synergies and opportunities for economies of scope seemed large (Suzuki and Kodama, 2004). MITI pursued this at a national scale. Since no other country explicitly targeted spillover, it was not surprising that Japan is where niche markets were most valued.

Niche markets were crucial when prices had not yet dropped enough to appeal to mass markets. Niches provided smaller markets with higher willingness to pay, which allowed PV producers to target real applications, with real customers, and often at prices without subsidies. Consumer electronics firms were powerful players in Japan's economy and became interested in PV as a way to differentiate their products.

Eighty percent of Japan's R&D in the 1980s focused on thin film silicon. This was significant, given their use for watches, calculators, and toys, even though the efficiencies were not high enough for thin film to serve mass markets. Ultimately, however, niche markets were insufficiently large for firms to expand and target economies of scale.

Perhaps the biggest contribution that Japanese made in policy was by combining technology-push and demand-pull (Nemet, 2009). While technology-push in the form of R&D had been strong and consistent since the 1974 Sunshine Project, in the 1990s Japan got serious on the demand side by subsidizing installations and supporting that by allowing net metering. They initiated a virtuous cycle of technology-push and demand-pull (Watanabe et al., 2000), which Germany subsequently imitated. This stimulated the world's first large-scale manufacturing of PV. The scale up by Sharp in the early 2000s was the most impressive and unlike any other Japanese firm. They showed it could be done by transforming a batch process to one that used continuous processing at an industrial scale.

Japan industrialized PV production with a sequence of R&D, niche markets, creation of markets, and by enabling scale up. Japanese solar firms became the largest in the world (Honda, 2008): "The Japanese industry forms the backbone of the global PV industry" (Foster et al., 2009). But Japan, and its top firm Sharp, lost its lead after 2005. Several explanations played a role: Japan did not anticipate that Germany's market would grow so quickly; the labor costs in Japan made it uncompetitive; they excessively emphasized quality; they had a slow response to the silicon price spike; and they over-invested in thin film.

Japan also contributed to the rise of Germany and later of China by making novel contributions to PV. Some of these, like serving new niche markets, improved the technology so that later entrants could adopt designs and production technology that was more advanced. Others—such as the rooftop subsidy program and demonstrating the cost-reducing benefits of scaling up—provided replicable models that could be deployed in other places with different configurations of resources and capabilities.

Post-war until oil crisis: industrial policy, research, lighthouses, and satellites

PV's origins in Japan date to the 1940s and the traumatic post-war period when providing subsistence in the form of housing and food were the central aims of the occupied government that emerged. As rebuilding progressed and the immediate needs of survival were met, inflation threatened to thwart progress. In this context, the Ministry of International Trade and Industry (MITI) was established in 1949 to coordinate Japan's international trade policy with other government agencies. MITI played a central role in the PV industry that emerged in Japan in the 1980s and 1990s.

MITI influenced exports, imports, and investment in all of the Japanese industries that are not specifically governed by other ministries. This breadth of influence allowed MITI to streamline policies that commonly conflict, such as pollution control and international trade. MITI functioned as the engineer of Japan's industrial policy, serving the dual roles of industrial mediator and regulator. A symbiotic relationship between MITI and Japanese industry created trade policies that typically complemented the ministry's work in improving domestic manufacturing. MITI is at least partially responsible for the development of most major industries in Japan, protecting them from import competition, aiding technological development, and facilitating mergers. MITI's authority was such that until the 1980s, it was the norm that prospective prime ministers were expected to have served a term as head of the ministry before taking the helm of Japan's federal government.

MITI's influence was strongest in the 1950s and 1960s, however conflicts arose as it navigated the dual responsibilities of promoting open markets and supporting domestic industries. By the 1980s MITI's policies began to change as a result of Japan's export success, which had caused tension with its trading partners. For example, MITI mediated talks between Japanese auto-makers and US policymakers who wanted to reduce Japanese automobile imports. The ministry's power waned further when it lost control over Japan's foreign exchange rate. In response to pressure, MITI began to loosen its import policies, drawing on its close ties to industry to encourage open markets in Japan. The liberalization of Japanese import policy caused an influx of foreign firms into the country. MITI's significance within the Japanese industrial world continued to decline until 2001 when it was dissolved and reorganized as the Ministry of Economy, Trade, and Industry (METI).

Sharp is the other key Japanese PV organization that emerged in the 1945–1973 period. Sharp has had the longest continuous involvement of any company in the solar industry. In 1915, founder Tokuji Hayakawa brought Sharp into the world of industrial manufacturing by launching the "Ever-Sharp" mechanical pencil. By the time it began researching solar cells in the late 1950s, Sharp was already experienced in mass-production, but there was a sense at the company that it was lagging behind other companies who were developing semi-conductor production. Sharp's over six decades of contributions to the PV industry stem in part from Hayakawa who in his 1970 biography said:

> I believe the biggest issue of the future is the accumulation and storage of solar heat and light … while all living things enjoy the blessings of the sun, we have to rely on electricity from power stations. With magnificent heat and light streaming down on us, we must think of ways of using those blessings. This is where solar cells come in. (Jones and Loubane, 2012)

Hayakawa's interest in solar was not only due to this vision, but also because he expected progress in PV to provide spillover benefits to other parts of his company. This systemic approach to PV was representative of Japan's national innovation system, as MITI also sought to maximize knowledge spillovers across industries (Sugiyama, 2018).

Sharp began PV research soon after Bell Labs launched its first efficient cell in the US in 1954 (Kimura and Suzuki, 2006). Sharp was among the first Japanese companies in PV along with NEC, Mitsubishi, and Sanyo entering just prior to it. As a result of this early start, Sharp's PV technology was based on the original Bell Labs design rather than the designs for space applications that later entrants adopted (Green, 2005). Further, unlike the others, Sharp did not have a semi-conductor business at the time so it started without the advantages of process know-how but with a cleaner slate. Their cell design was simple and the p–n junction discussed in Chapter 3 was well documented in the literature (Chapin et al., 1954), so Sharp did not face difficulties in developing its own cell and did not need to license the technology from Bell Labs as the first US PV companies, such as Hoffman, did. After only a couple of years of PV research Sharp installed its first commercial module, a 70W installation, at Tohoku Electric Power's communications relay station in November 1958. Like other firms entering PV, important initial applications of photoelectric cells were as light detectors that could be used in electronics.

Sharp began processing mono-crystalline cells in 1959 and began producing solar cells in large quantities in 1963 at its Nara plant. Sharp wasted no time in demonstrating its product. That same year they had begun mass-manufacturing, and they used 23W of solar cells to power the No. 1 Tsurumi light buoy near Yokohama Port. Sharp piloted novel uses for this new power source including lighthouses, navigation buoys for the Japanese coast guard, and satellites for the Japanese space program. In 1966, Sharp installed a 225-watt solar power system on a lighthouse on Ogami Island in Nagasaki Prefecture, the world's largest power system of its type at the time. Subsequently, Sharp installed solar battery systems on 226 lighthouses in Japan from 1961 to 1972 (Perlin, 1999).

A new market emerged when the Japanese Scientific Satellite Program began in 1965. In 1967, Sharp began development of solar cells to be installed on satellites and in 1974 it produced the first Japanese PV module for space, which was installed on the Japanese space program's Ume satellite launched in 1976. By the early 1980s, Sharp had become the exclusive supplier to the Japanese National Aeronautics and Space Development Agency.

Sunshine Project, July 1974–1984

Much as it did in the US, the Arab Oil Embargo of October 1973 catalyzed energy policymaking in Japan. The Arab nations blocked oil shipments to Japan as well as to Canada, the Netherlands, the UK, and the US. The oil price shocks hit Japan especially hard because it had almost no domestic energy resources and thus, the sense of crisis was deeper in Japan than even the US. Another similarity to the US was that Japan's comprehensive reaction to the crisis was enhanced by precursor efforts that became available to scale up once the crisis mobilized public concern and resources. By Fall 1973, the Agency of Industrial Science and Technology (AIST) within MITI had already set up the Large Scale Industrial Program as its main industrial technology policy program. From that program, MITI formally launched the Sunshine Project in July 1974—on the heels of President Nixon's Project Independence, but with a different focus. It targeted six R&D areas, including solar and coal, but notably not initially including nuclear power.

The goals of the Sunshine Project were in general to address the vulnerability of the Japanese energy supply infrastructure, and specifically to reduce Japan's dependence on foreign oil. The program focused on innovation, emphasizing the:

> Development of a new technology to produce clean energy by the year 2000 through various medium-term research projects of several years each, and to satisfy a great percentage of the energy demand for several decades in the future. (Matsumoto, 2005)

The project was distinct in that even beyond energy it was Japan's first "large scale long term technical development effort which deserves to be called a national project" (Kashiwagi, 1996). Its technical focus evolved, from an early priority on solar thermal electricity to PV in 1981 and thence to amorphous silicon PV. This latter focus was transformative in that it opened up the consumer electronics market and entrained some of Japan's largest corporations to become active in PV technology commercialization.

As in the US, awareness of energy issues in Japan had been building before the crisis. In the early 1970s Japanese industrialists had called for the government to do something about the country's heavy dependence on foreign oil, on which these companies depended. Japan had few domestic energy resources so plans were made to consider increasing hydroelectric plants, nuclear power, and liquefied natural gas. Of these, nuclear was the most attractive option. Even though Japan had begun building nuclear plants in the late 1960s, it was still not in favor politically, due to concerns about storage of waste and the potential for accidents. Originally, solar could not compete with any of these and thus was not a priority technology program. It was clear, however, that the solar resource was extremely large, but it was just too expensive. Japan needed a project to develop solar into a viable response to its dependence on imported oil.

The ideas for the Sunshine Project originated in a government laboratory, the national Electrotechnical Laboratory (ETL), in the spring and summer before the oil crisis. In May 1973, MITI called for the establishment of a department of "new energy" and in August MITI began to ask agencies, including ETL about "new energy" technologies (Matsumoto, 2005). Within ETL, Takashi Horigome, a lab manager, had become a vocal public advocate for PV, pushing the notion of solar in the public sphere, in the press and on TV. He began to propose the idea for a research initiative in PV to AIST at the beginning of 1973. Still, the ETL leadership was deeply skeptical:

> One of Horigome's assistants said that the lab felt as though something like solar energy will probably never amount to anything. And allowing that kind of childish, toy-like research to be carried out in our research laboratories, and assigning research funding for something like special research at ETL? Absolutely outrageous. That was the kind of attitude they had at the time. (Shimamoto, 2014b)

As described at the beginning of this chapter, Suzuki Ken, head of R&D at AIST had just been tasked with leading the solar program there. He quickly contacted Horigome directly to enlist his active participation. Suzuki was to become a strong supporter of Horigome, convincing ETL leaders to allow Horigome some space to do PV research. Once MITI's intentions became more apparent, Horigome and Suzuki began discussions about assembling a plan, which they sent to MITI for a Large-scale Industrial Technology Program in solar energy. The Minister of International Trade and Industry announced the plans for the program in August 1973. The Arab Oil Embargo began two months later.

With the oil crisis, MITI consolidated its position by taking the lead on an issue of national importance, on which there was broad support of acting. It set up the Office of Alternative Energy Policy with the task of developing new energy sources that could address the crisis. It set up the Agency of Industrial Science and Technology to develop technologies that could be useful to Japanese industry. It adopted a long-term orientation, across multiple technologies, at the scale of the recently completed Apollo Project and with the same high prioritization of Project Independence. Whereas most projects in Japan were expected to show a positive return on investment in five years, here MITI adopted a longer-term perspective. If it took 25 years to develop inexpensive PV, that was acceptable since the issue of energy dependence was so important to the country.

Because the initial calls for the government to do something about energy came from industry, the motivation for the Sunshine Project was about supporting industrial policy. Rapidly growing companies needed reliable access to energy to continue to grow. Solar was of interest in that it might one day enable that growth, not because solar could become an industry of its own. Similarly, to the US, these precursor efforts were able to be quickly implemented and expanded once the policy window of the Arab Oil Embargo opened. They become the

Sunshine project after the crisis hit. It officially began in July 1974 and was headquartered in MITI.

Under the Sunshine Project research would be performed by universities, research laboratories, and Sunshine Project headquarters at the Agency for Industrial Science and Technology (AIST). Technology development, including demonstrations, was managed by a new entity, the New Energy Development Organization (NEDO) launched in 1980, which worked closely with industrial partners, such as Kyocera, Sanyo, and Sharp. The program set up an annual meeting, the Japanese PV Science and Engineering Conference (PVSEC). AIST officials decided to give the project a name, Sunshine, that would engender public support in the context of large required budgets and perhaps three decades of effort.

Even though solar was the initial focus, Suzuki felt the need to hedge against the risk that solar technology would fail and sought additional energy programs to include. The final technological scope included four programs: 1) solar energy 2) geothermal energy 3) gasification and liquefaction of coal, and 4) hydrogen energy. Notably, nuclear fission was not included, in part because strong anti-nuclear sentiment had existed in Japan since 1945. There was a feeling that nuclear would eventually become used in Japan but, in contrast to other countries, only after solar became widely adopted.

Under the solar energy category, solar thermal electricity (STE) received the most funding because it was thought to be closer to commercial readiness. However, it turned out that due to high moisture in Japanese air, sunlight there was too diffuse for effective STE. As a result, by 1980 PV became the focal point of the solar program. This shift to PV also affected the types of projects being funded. Instead of large STE demonstrations, PV applications in contrast were targeted to small scale, such as tunnel lighting, satellites, and farming. Expectations were that it would take a long time for PV to become affordable. Cost reductions were a central thrust of the R&D program and its goal was to reduce the cost of PV by a factor of one hundred. For example, silicon was expensive, so a big part of early efforts was to reduce silicon cost and the amount needed for a cell.

Because its budget was negotiated at the height of the oil crisis, the project was fully funded and the first-year budget for the Sunshine Project 1974, 2.3 billion yen ($20 million), was taken from the general account for solar (Shimamoto, 2014a). Its budget rose by a factor of four by 1978. Solar energy was one of six R&D focus areas, about 20% of a total $150 million per year budget, most of which was for capital equipment and procurement. Companies and universities were expected to pay their own labor costs with the exception of ETL, which used its allocation of Sunshine budgets to pay its staff. These budgets were large and stable throughout the 1970s. To justify the large budget, the project proposed ambitious goals, with an extended time frame of 2000. Despite the multi-pronged technology plan, solar was the dominant target. The projections are dramatic. Half of energy would come from solar energy by 1990, and about two thirds by 1995.

The Sunshine Project had strong corporate involvement from its inception. In one sense, companies were the primary initiators, calling for a new energy strategy well before the 1973 crisis. Hitachi, Matsushita, NEC, Toshiba, Toyo Silicon, and Sharp were all early entrants in the PV program, as each had done some R&D in the area. But at the time Sunshine started, Sharp had the most experience. These companies saw the program as a way to improve their technology capabilities. Not only did Sunshine aid in the understanding of the properties of photoelectric materials, they also made the firms more engaged in developing the technology. These firms were perhaps most influenced by the emerging notion that there would one day be a substantial market opportunity for PV because of the ambitious targets set in the program, for example nearly half of energy by 1990 (Kimura and Suzuki, 2006). In contrast, the newly available R&D funding was less of a driver. One indication that expected markets were more important than available public R&D funds was that these firms, and later the important players Sanyo and Kyocera, began to invest their own R&D funds in PV (Watanabe et al., 2000).

Sharp established its Sunshine Project Promotion Department in 1974 and appointed as its head Kimura Kenjiro, who had led PV manufacturing at Sharp since the 1950s. It researched the standardization of solar cell production, an effort to make the process more suitable to automation. It was particularly interested in the edge-fed growth process from Tyco in the US, which Kyocera had adopted. This avoided the losses of expensive silicon when ingots were cut. It is important to realize that there were almost no production facilities in Japan at the time of the Sunshine Project. Some US manufacturing equipment became used at this time but there it was not feasible to reduce costs with US-made equipment. So domestic semi-conductor equipment enterprises and PV companies independently developed dedicated production equipment for themselves. Each company has created its own individual equipment for its production lines. There was no common specification for mass production facilities that would allow one firm's equipment to work with another's. In contrast, later on Germany developed third party suppliers of production equipment that could be adopted by several companies and could be combined with other equipment suppliers' machines to create a line. This enabled a degree of specialization, which the Japanese in-house model could not attain.

Sharp led the way in commercializing PV, introducing the world's first solar calculator in 1976, and three years later, demonstrating a hybrid house powered and heated by solar panels. None of these ventures proved profitable for Sharp—the demonstration projects supported by NEDO were too small. But leadership continued to invest perhaps in part due to the enthusiasm of the founder. Kyocera followed a similar path, beginning its own R&D in 1975, starting production in 1979, and beginning mass production in 1980. Still, there were limits to firms' enthusiasm. Profits were a distant prospect and many firms were reluctant to take on PV development projects because they were not confident they could meet Sunshine's ambitious targets—even after 10 years.

By the end of the 1970s, Japanese firms began to make major gains in producing cells. One major accomplishment was a collaborative 500 kW module plant with a substantial amount of automation. At least nine companies, including the largest

PV producers in Japan, contributed production steps to the plant. This integrated production facility was the closest thing Japan had to the US Block Buy program, which was in process concurrently. Both had the same goal—to shift PV manufacturing from a labor-intensive customized process to a more automated mass-production oriented system. Construction began in 1980 and the plant was operational from 1982 to 1985 (Shimamoto, 2014a).

The most dramatic technical breakthrough however was in developing amorphous silicon (a-Si), a type of thin film PV. Research on a-Si had begun in the 1950s at RCA in the US and later developed by Stan Ovshinsky and Energy Conversion Devices in the 1960s. A unique feature of the a-Si program in Japan was that universities were deeply involved, which was not the case with traditional crystallized silicon (SI) projects. A-Si had the promise of becoming much cheaper than crystals because only a small amount of silicon, which was then expensive, was needed to deposit on a plastic substrate. Entire steps such as growing crystals and slicing wafers could be avoided. Their backing was flexible allowing them to be easily shaped to contoured surfaces and they worked well in indoor environments with fluorescent lighting. Their efficiencies were much lower, however, by roughly a factor of two. If the same efficiency gains that x-Si had seen could be replicated in a-Si, then it was a no brainer that a-Si would be the winning technology. Companies bet big on a-Si, in the late 1970s but also later in the 2000s (Endo, 2017). As it turned out the efficiency gains in a-Si never materialized, which was a contributor to Japan giving up its worldwide lead in PV production.

Second oil crisis in 1979 intensifies the Sunshine Project

The Iranian Revolution in early 1979, and the oil crisis it precipitated late that year, reinvigorated the Sunshine Project. The most important change was the launch of a new agency, the New Energy Development Organization (NEDO) in 1980. Its role was to coordinate research with three new foci: practical applications, manufacturing improvements, and cost reductions. Corporate involvement was much more substantial in this period as early commercial opportunities became more apparent than at the outset of the Sunshine Project. However, solar still had not emerged as an industry in its own right. An inconvenient truth about this next phase of the Sunshine Project was that over half of the budget in the 1980s went to coal technologies.

The Law for the Development and Promotion of Oil-substitute Energy (Law No. 71) authorized NEDO in May 1980. Takashi Horigome, a former Leader of Power Research Division of the National Electrotechnical Laboratory became head of NEDO's solar energy office. Steady funding also emerged in this period via the Special Account for Alternative Energy Development. Most of the budget of the Sunshine Project after 1980 came from this account, which was funded via a tax on electricity and coal purchases. Japanese R&D funding was remarkably stable from 1980 until 2003, in contrast to volatile budgets in the US (Figure 5.1). Japan's progress in this period was such that there was a growing concern in the US that Japan was about to make major advances in PV technology, particularly in a-Si,

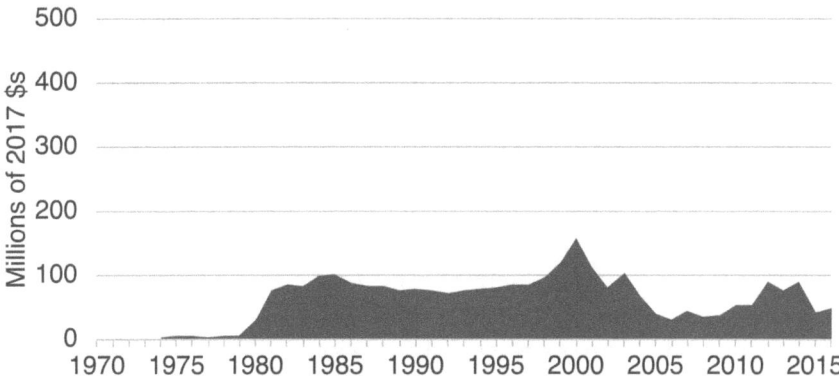

FIGURE 5.1 Japanese solar PV funding (millions of 2017$s). For comparison, the vertical scale is the same as for Figure 4.2 on US funding.

that would leave the US behind as it was cutting its R&D budgets (Maycock et al., 1983). Motivated by these concerns, the US Department of Energy sent a delegation to Japan in 1983 to assess the situation there (Shimada, 1983).

The early 1980s also saw the first pilot manufacturing plants that aimed to do what was then considered "mass production." As had happened in the late 1970s in the US, and to a smaller extent in Japan then too, government procurement was to be a big part of the mid-1980s scale up. There was thus a focus on standardization of designs and of assembly steps. This process provided opportunities for producers to accumulate experience in production and find ways to adjust their processes through learning by doing. Electricity prices in Japan were approximately twice those of the US, so cost reduction targets did not need to be as ambitious.

Japanese corporations took on a much more active role in PV development from 1980 onwards. NEDO made conscious efforts to create links between industry, government, and academics. It sought to establish a dialogue of trust, through the encouragement of participants in the consortia to provide feedback to each other on research and development progress. NEDO had to navigate a tricky tension in working out ways for rival firms to cooperate on research (Matsumoto, 2005). This tension heightened as profitable commercialization became imminent.

The most important mechanism for encouraging competitors to collaborate came from technology rather than policy. The substantial progress in a-Si technology fueled the interest of consumer electronic firms in commercial applications of PV. For example, Sanyo began to mass produce a-Si in 1979 and remained active in consumer electronics PV throughout the 1980s. Sharp began work on integrating small solar cells in its calculators in 1976, marketing the first solar cell calculator in 1980. To be sure policy had a role in supporting a-Si research and the government's efforts to create a research consortium made participants aware quickly of the progress being made. The main driver of firms' interest, however, was unrelated to addressing import dependence. Firms were mostly concerned with the emerging commercial opportunities from the inclusion of a-Si technology in their own consumer electronics businesses, which remained their core concern. Calculators and electronics were easy to reverse engineer so knowledge spread quickly. In 1982, 80% of Japan's 1.7 MW annual production was for consumer electronics (Maycock et al., 1983).

The government's goals under the Sunshine Project were not entirely aligned with these developments; they wanted firms to progress toward manufacturing larger solar modules for homes and businesses. But the firms were powerful and PV was still not inexpensive enough to compete with grid power, thus consumer niche markets continued to dominate. So even as NEDO expressed concern that solar in consumer electronics would not help oil imports, a consensus emerged was that this direction should be tolerated because it was assumed these niche applications would eventually spillover to large-scale power generation. This acknowledgement and encouragement of technology spillovers was a hallmark of NEDO and of MITI in general.

The 1980s saw a divergence emerge between firms pursuing a-Si and those developing the more traditional crystallized silicon PV. The latter group included Hitachi, Toshiba, and NEC. Rather than difference over research directions, this divergence involved substantial financial commitments in building production facilities. Hitachi set up a 250 kW per year line and NEC set up a 100 kW per year line to produce cells in order to support its mainline business of communications equipment, like using solar to power remote telecom repeater stations and satellites. Toshiba focused on a novel type of crystalline PV, using "ribbon silicon," and worked with NEC and Hitachi on a 500 kW per year plant in 1983. The ribbon silicon process had the advantage of not requiring slicing since it requires that the silicon be pulled to the preferred thickness, saving money and silicon in the process. Kyocera, which later would become the largest producer, created a joint venture with Mobil and Matsushita to develop ribbon silicon called Japan Solar Energy Company (JSEC). Matsushita also worked on compound cells, using stacks of more than one p–n junction.

Meanwhile, Sanyo Electric, Fuji Electric, and Mitsubishi Electric developed a-Si to focus on consumer electronics. Of these, Sanyo was the most important player in that it commercialized a-Si for calculators in the 1980s and became so successful at it that it was able to sell its cells to other large consumer electronics companies such as Casio for their own solar calculators. Sanyo was serious in its commitment to PV—it funded its work with its own R&D funds rather than its competitors, which used Sunshine funding. Sharp remained neutral on the a-Si/x-Si divide in the early 1980s by developing both technologies.

A lull post-Sunshine 1984–94: demonstration projects and small markets

Like Project Independence in the US, the Sunshine Project eventually faced declining political support as the urgency created by the 1970s oil crises faded. Oil prices declined in the early 1980s and crashed in Fall 1985. Even though the R&D funding levels were maintained thanks to the Special Account, corporate participation declined in the mid-1980s. Aside from consumer electronics niche applications, publicly funded demonstration projects were essentially the only market in the 1980s (Kimura and Suzuki, 2006). Many firms withdrew from PV altogether because Sunshine-funded demonstration markets were too small and the consumer electronics markets became saturated. There was concern that a valley of death existed in that the basic technology was well advanced and in the public domain, but the production technology that resided at the firms was proprietary and not making as much progress. It would take until the rooftop program in the mid-1990s to overcome this gap.

Still, the consistency of R&D funding made Japan's experience different from that of the US. Scientists and engineers stayed involved in part because the R&D funding continued. Not all companies divested of PV; Kyocera, Sanyo, and Sharp all stuck around. The NEDO program was successful in its cost reducing goals as prices declined by a factor of four to five from 1980 to 1985.

When global interest in solar declined through the 1980s, Sharp continued to serve as a bastion for the production of solar cells by introducing a new line of solar-powered calculators. In 1983 it was producing PV for applications in maritime safety, weather stations, agriculture, and other off-grid applications. In 1984, in Shinjo, Japan, Sharp opened its first manufacturing plant devoted entirely to solar power-related products. It bought production equipment from Energy Conversion Devices in the US that could produce 3 MW of a-Si annually. Further, Sharp continued solar research and development when others were not, developing new solar technologies and applications. One reason offered as to why, even in the face of rapidly falling oil prices, Japan continued to invest in solar R&D is that the 1970s crises were so much more acute in Japan than elsewhere. The memories of them, and the possibility of recurrence, lingered.

While the period from 1984 to 1994 was mostly a lull in PV efforts around the world, it was also a period in which ideas emerged for how PV could begin to serve markets other than niche applications and government sponsored demonstration projects. First, The Japan Photovoltaic Energy Association (JPEA) started in the late-1980s to provide nascent legitimacy for the notion that PV could be an industry on its own, rather than a minor input to established industries. Firms such as Sharp and Sanyo were already so closely involved with MITI through their world-leading consumer electronics businesses that they naturally collaborated with MITI in developing ideas for a rooftop subsidy program. This corporatist governance approach excluded new entrants and enhanced the influence of established domestic conglomerates in policy design. These firms had been advocating since the 1980s that larger markets were needed in order for costs to come down. These arguments became more prominent as the industry matured and a collective voice in support of market support coalesced.

Second, a MITI Energy outlook in 1990 targeted a high share of renewables in the 10–20-year time horizon. Third, and also in 1990, the PV Power Generation Technology Research Association (PVTEC) was launched as an R&D Consortium focused on identifying opportunities for cross-technology spillovers (Watanabe et al., 2000). Fourth, Japanese utilities surprisingly started to offer voluntary net metering in 1992. This change allowed residential customers to sell back to the utility any excess PV power from their rooftop systems, which made their economics much more favorable. Utilities had strongly opposed such a scheme in the 1980s. One interpretation for the change is that MITI's renewables targets made utilities realize they had to get on board. More cynically, being proactive meant that they might be able to avoid regulatory action and thus could withdraw the net metering on their own.

Fifth, international collaboration was building in this period. Japan was highly aware of developments in Europe. It had seen Project Megawatt by Alpha Real in Switzerland and the first market creation policy was launched in Germany in 1990. Japan also entered into bilateral agreements: with Australia on field testing and remote applications, with the US in exchanging professionals for information exchange, and with the International Energy Agency (Takahashi, 1989).

Finally, climate change emerged as an issue in Japan in the early 1990s with the implementing agreement for international cooperation signed in 1997 at the Kyoto International Conference Center. This new policy direction raised expectations about the deployment of renewables. It provided a new driver for PV, distinct from industrial policy and energy independence that had bolstered the Sunshine Project.

It is perhaps indicative of all of these drivers that in the 1990–91 period with the disastrous collapse of the Japanese bubble economy, nearly every government program was cut, but clean energy was popular and solar R&D funding was maintained.

Rooftop subsidy program 1994–2005

Two important precursor programs set the stage for the most important Japanese PV policy since the Sunshine Project, the rooftop subsidy program. First, the 700 Roofs Program launched in late 1993 and gave MITI the first real indication of how much latent demand for PV existed. Over 1,000 applicants requested the subsidy. MITI expanded the program's budget to 1,000 roofs in 1994 and received 5,000 applications. It became clear that a major subsidy program would be popular. Second, after lobbying by the Japanese Photovoltaic Energy Association (JPEA), the "Guidelines for the technical requirements for PV grid connections with reverse flow" were adopted in 1993 (Tatsuta, 1996). While utilities had voluntarily offered net metering a year earlier, this package of policies reduced the transactions costs associated with installing a system (Shimamoto, 2014a). They provided confidence to PV adopters that they would be able to continue to sell excess power back to the grid. Its legal implementation removed the possibility that utilities could withdraw the offer.

Despite the general enthusiasm for PV and the robust set of drivers that had been building, passage of the rooftop subsidy program faced challenges (Kimura and Suzuki, 2006). First, there was no precedent for providing a subsidy to individual residences. Second, the increasingly influential Finance Ministry was dismissive of subsidy programs in general. Third, the program was risky in that it depended on individual households choosing to adopt solar and being willing to invest thousands of dollars on their own. Developing the industry and catalyzing cost reductions would not happen if households were unresponsive to the program.

MITI overcame these challenges as a result of four positive drivers (Kimura and Suzuki, 2006). First, there was a broad consensus that the subsidy program would stimulate learning by doing and lead to cost reductions according to the learning curve. Second, as a result of the drivers described above, MITI and NEDO felt strong pressure to do something substantial and highly visible to increase the deployment of renewables. Third, the 700 and 1,000-roof pilot programs bolstered confidence that homeowners would enthusiastically pay for a PV system. Fourth, the taxes on electricity and coal had adequately funded the Special Account for Alternative Energy Development so funds were available to disburse quickly.

These drivers led MITI to launch the first major market creation policy in the world. Nearly simultaneously, in April 1993, the Agency of Industrial Science and Technology launched the New Sunshine Project, which renewed R&D funding for PV and set long-term goals of cutting carbon dioxide emissions by 50% by 2030 and having solar provide one third of energy supply by 2030.

MITI launched the rooftop subsidy program in 1994. While the central aspect of the program was a rebate for a portion of the cost of installations, it included a set of complementary policies. It simplified permitting procedures for installations, it provided technical guidelines for connecting systems to the grid, and supported the net metering scheme that the utilities had voluntarily adopted.

The rebate scheme itself originally provided adopters of residential PV with a cash grant of 50% of the full cost of installing a system, including hardware, installation, wiring, etc. Consistent with the notion that the learning curve would lower costs, the scheme also specified that the rebates would decline over time and eventually end after ten years. This notion of declining rebates was perhaps Japan's biggest policy innovation and would be replicated around the world. Rebates were set at 50% for 1994–96, at 33% for 1997–99, and thereafter at a fixed yen per W of capacity: JPY150–270 per kW in 2000, 120 per W in 2001, 100 per W in 2002, 90 per W in 2003, 45 per W in 2004, and 20 per W in 2005. Figure 5.2 puts these rebates in consistent terms using the average prices of installed systems. One can see both that the subsidies declined in constant $ per W terms quite rapidly.

After three years of success at a level of 1,000 or so houses per year, MITI substantially enlarged the program in 1997, increasing the budget by a factor of three to four from 1997 to 1999. By 2000, Japan was the largest PV market in the world. It became obvious, perhaps for the first time, that firms could make money in solar in the near term. Installations grew from 539 in 1994 to 15,879 in 2000 to 54,475 in 2004. Installations were limited by the program budget and the program was over-subscribed in every year—by a factor of five in the early years. Eventually political support for the project waned. It was almost eliminated in 2000 as powerful entities such as electric utilities and the nuclear industry began to view PV not as just a novelty, but as a threat. In fact, the Federation of Electric Power Companies of Japan, the trade group for electric utilities, had paid for monthly front-page ads denigrating renewable energy and championing nuclear power in the

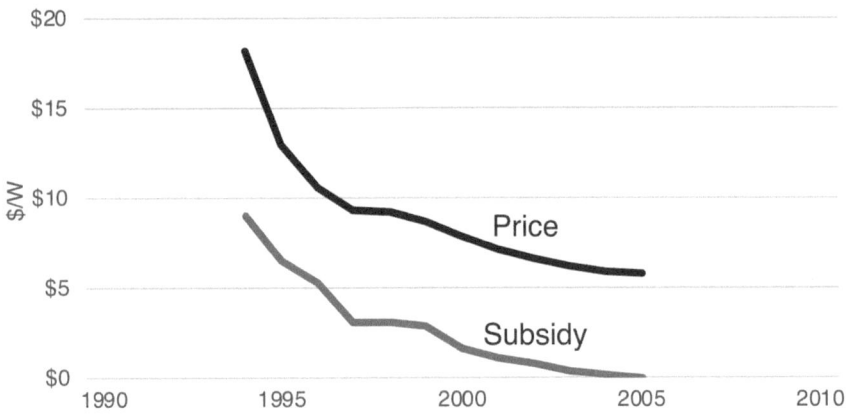

FIGURE 5.2 Prices of installed systems in Japan from 1994 to 2005.

1990s. The subsidy budget was cut by over half in 2003 and by 2005, it ended. At that point no subsidies were available, although net metering remained.

Over 200,000 systems and 0.8 GW of capacity were installed over the 11 years of the program, drawing on a total budget of $1.1 billion. The program provided real evidence for the existence of a PV learning curve. Prices of installed systems declined by more than a factor of three over the course of the program. The combination of the R&D funding from the New Sunshine Project and the rebate program catalyzed a "virtuous cycle" that led companies to invest their own R&D funding to improve PV (Watanabe et al., 2000). That outcome is quite clear even if it is harder to say if it was an intentional aspect of designing the program. Thus, the most dramatic outcome was the emergence of a serious PV industry in Japan. As Figure 5.3 shows, the global PV market went from one led by niche applications before the rooftop program to one where rooftop applications dominated (Maycock, 2005). With the Japanese rooftop program, the most important market for PV was no longer niche applications but consumer electricity; it was now competing directly with grid electricity.

Sharp scales up

Kyocera, Sanyo, and Sharp all benefited from the rooftop subsidy program. The first two were arguably more active in PV in the 1980s and early 1990s, but Sharp went beyond them and became the largest PV producer in the world. From 1994 to 2005, it scaled up by a factor of 200. In 2005 its plants had 0.5 GW of production capacity. Its global market share in 1993 was less than 2% and by 2005, at the end of the subsidy program, was 28%. The 30% market share it reached in 2003 is the highest market share any PV company has had since comprehensive global data first became available in 1988. None of the Chinese giants that have emerged in the past decade ever exceeded 10%. Sharp scaled up smoothly and at

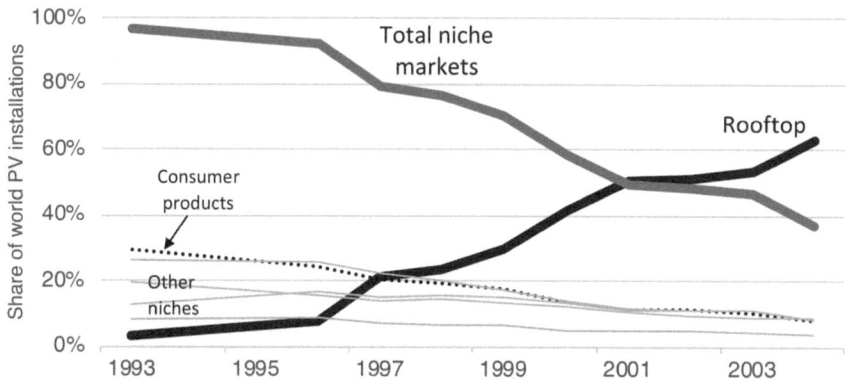

FIGURE 5.3 Share of world PV market by application area.

the right time. One of the intriguing aspects of Sharp, as well as the other Japanese producers, was that they never published data on their PV production costs because PV was just a small part of a giant conglomerate. That opacity was to change dramatically with the pure-PV companies that got started in Germany and China around the millennium.

Sharp began to expand internationally in 2003, building a 40 MW line in Memphis, Tennessee to manufacture commercial and residential solar for the North American market. In 2004, Sharp set up a 40 MW line in the UK to better serve the European market. Having dominated the crystalline silicon market, Sharp set its eyes on applications of thin film solar cells. In 2008, Sharp increased its production capacity for thin film solar cells from 15 MW per year to 160 MW per year at its plant in Katsuragi City. In 2010, Sharp doubled its thin film production capacity with the opening of a new plant in Sakai City. Sharp researchers and engineers rarely interacted with researchers and engineers in European and American research institutions and companies. At most, they attended academic conferences.

However, 2005 was a high-water mark for Sharp's leadership of the global PV industry. Its market share shrank to less than 10% by 2007 and continued to decline thereafter to the low single digits. The Japanese industry still survives today but is much smaller and has re-focused on niche high efficiency products. One wonders how Sharp could have fallen so far so quickly from such a dominant position. It knew first-hand the benefits that come from increasing scale, it was close to the supply chain in Japan, it had established an international presence to take advantage of multiple markets, and its decades of R&D and experience in the industry gave it a knowledge stock that no other company could even approach.

Several possible explanations for why Sharp relinquished its lead emerge out of interviews in Japan and elsewhere. First, Japanese firms, including Sharp, did not anticipate that the market in Germany would grow so large so quickly. Perhaps the insularity described above contributed (Harborne and Hendry, 2012). They were left flat-footed. Moreover, they were not able to respond once it became obvious that growth was so fast since they were used to producing for their own domestic market, which had dominated global PV demand for 10 years. As one interviewee put it, "we were not paying much attention to Germany." Sharp had little experience selling in Europe and partnering with European distributors. From 2004 onwards, almost all of the demand was in Europe as Japan ended its subsidy program in 2005.

Second, Japan was an inherently higher cost location than those of its emerging competitors, particularly in China. Salaries for personnel, material costs, transportation costs, and electricity were all much more expensive in Japan than in China. Japan's emphasis on environmental protection also raised the costs of operations compared to businesses in China, where environmental concerns in the specific zones in which PV companies were operating were often given a pass by local authorities. China had reliable electricity, which was essential for processes such as crystal pulling in which temperatures need to be constant for 36 hours. Japanese firms were reluctant to move production to low cost locations due to concern about their intellectual property leaking to rivals. Eventually, Sharp did pursue overseas production in lower cost locations, but due to its much more circumspect and languid decision-making processes, those efforts were too little too late.

A related reason was that Sharp took pride in its quality-oriented production, perhaps a cultural spillover from Japan's success in the automobile industry. It did not make the same cost-saving production compromises that China's firms continually targeted. Sharp also interpreted the rapid growth in the PV market since 2004 as a reason that it could continue to sell at high prices because it assumed that demand for quality would always exist. Sharp sought to maintain its advantage through high efficiency, durability, low degradation, and other quality related advantages. The disruptive manufacturing that emerged in China compromised on these attributes and made up for it with much larger reductions in cost. That did not happen in Japan.

Third, Sharp was not able to deftly adjust to the dynamics in the market for purified silicon. As one interviewee put it with understatement: "procurement of materials, particularly cheap raw material silicon procurement did not go well." Even as silicon prices rose by a factor of 10 from 2004 to 2008, incumbent silicon producers like Wacker, Hemlock, and Tokuyama were hesitant to expand production. They may have been skeptical that the demand for silicon was too dependent on potentially fickle policy makers. Alternatively, they may have been keen on maintaining prices well above $100 per kg with costs of production around $15 per kg. In any case, silicon producers began to assent to their customers and build new production facilities. In return, they wanted guarantees in the form of large prepayment demands and commitments to purchase years of production at high prices. Here, Sharp's position on solar, as part of a large established and successful conglomerate, became a liability. Start-up companies, such as Q-Cells in Germany and Suntech in China, were able to agree to these long-term contracts since they had no other options for accessing input materials. In contrast at Sharp, prepayments and long-term contracts were substantial enough that they could jeopardize the solvency of the entire company. Start-ups had no choice but to roll the dice with the silicon makers' demands, whereas Sharp's board refused to play. Its supply of input material was consequently squeezed just as its competitors were massively expanding. More generally, the decision process in a large, Japanese conglomerate was too slow compared to start-ups—and because of the design of the German feed-in tariff, the market demanded speed. A related issue is the number of veto points existing in a large corporation, whereas the Chinese, and to some extent German, start-ups were led by individuals who, by the nature of being entrepreneurs, were much more comfortable taking large risks.

Fourth, Sharp's more deliberate response to the silicon shortage was to massively expand into technology that used much less silicon, the thin film amorphous silicon technology that Japan had pioneered in the late 1970s and 1980s. It spent 100 billion yen ($1 billion) on an a-Si plant near its headquarters in Sakai City, Osaka to co-locate production of a-Si solar modules and liquid crystal display televisions, which was a much more strategic product for Sharp. Economies of scope was the goal for the plant that came on-line in 2010. However, the efficiency gains needed for a-Si to be competitive never materialized. As it turned out, commercial a-Si never attained anything close to the efficiency of crystalline silicon. Had a-Si efficiencies doubled to the range of x-Si, perhaps this big bet would have had a different outcome for Sharp solar. But instead, the Sakai City plant became a millstone on Sharp's neck, at least of the same size as expensive long-term silicon procurement contracts would have been. This was not the only occurrence in the history of PV in which a bet against crystallized silicon turned out badly, as we will see in Chapter 6.

Fifth, this period of 2004–08 was when technology leadership in the PV industry shifted from producing firms to equipment firms. As processes became less labor-intensive and more automated, the real high-tech aspect of PV was not in the products or in the way lines were configured but in the equipment that made up those lines. Because a-Si never became sufficiently efficient, Sharp's bet on a-Si exacerbated this emerging shift in value added to the machine makers.

Those in the PV industry in Germany, the equipment suppliers in particular, had been closely watching the development of the PV industry in Japan in the 1990s. They paid especially close attention to the rooftop subsidy program launched in 1994. It stimulated Japanese homeowners to invest approximately $1 billion in their own PV systems while electricity ratepayers and coal users helped fund those installations with about $1 billion of their own. It also helped support an industry that eventually attained a global leadership status, embodied by Sharp, not seen before or since. The policy design showed that a predictable schedule of payments helped create strong incentives for homeowners to invest while stimulating the industry to reduce costs as the subsidies were scheduled to fall and ultimately sunset. Two German policy entrepreneurs saw this as further evidence of what could be done, albeit on a scale two orders of magnitude larger.

Note

1 This paragraph is based on information in the definitive account of Japan's Sunshine Project, Professor Minoru Shimamoto's 2014 book, *National Project Management: The Development of Photovoltaic Power Generation System under the Sunshine Project*.

References

Chapin, D. M., Fuller, C. S. & Pearson, G. L. 1954. A new silicon p–n junction photocell for converting solar radiation into electrical power. *Journal of Applied Physics*, 25, 676–677.
Endo, E. 2017. Influence of resource allocation in PV technology development of Japan on the world and domestic PV market. *Electrical Engineering in Japan*, 198, 12–24.
Foster, R., Ghassemi, M. & Cota, A. 2009. *Solar Energy: Renewable Energy and the Environment*, Boca Raton, FL, CRC Press.
Freeman, C. 1987. *Technology Policy and Economics Performance: Lessons from Japan*, London, Pinter.
Green, M. A. 2005. Silicon photovoltaic modules: a brief history of the first 50 years. *Progress in Photovoltaics: Research and Applications*, 13, 447–455
Harborne, P. & Hendry, C. 2012. Commercialising new energy technologies: Failure of the Japanese machine? *Technology Analysis & Strategic Management*, 24, 497–510.
Honda, J. 2007. History of Photovoltaic Industry Development in Japan. *Proceedings of ISES World Congress 2007* (Vol. I–Vol. V). Springer, 118–123.
Johnson, C. 1982. *MITI and the Japanese Miracle: The Growth of Industrial Policy: 1925–1975*, Stanford, CA, Stanford University Press.
Jones, G. & Loubane, B. 2012. *Power from Sunshine: A Business History of Solar Energy*, Cambridge, MA, Harvard Business School.
Kashiwagi, T. 1996. Technological breakthrough and global cooperation. In: Suzuki, Y., Ueta, K. & Mori, S. *Global Environmental Security: From Protection to Prevention*. Berlin, Springer.
Kimura, O. & Suzuki, T. 2006. 30 years of solar energy development in Japan: Co-evolution process of technology, policies, and the market. Berlin Conference on the Human Dimensions of Global Environmental Change: "Resource Policies: Effectiveness, Efficiency, and Equity", 17–18 November 2006. Berlin, 17–18.
Matsumoto, M. 2005. The uncertain but crucial relationship between a 'new energy' technology and global environmental problems. *Social Studies of Science*, 35, 623–651.
Maycock, P. D. 2005. *PV Technology, Performance, Cost 1995–2010*. Williamsburg, VA, PV Energy Systems.
Maycock, P. D., Shimada, K., Stirewalt, E. N. & Hunt, V. D. 1983. *America Challenged: Photovoltaics in Japan*. Williamsburg, VA, PV Energy Systems.
Nemet, G. F. 2009. Demand-pull, technology-push, and government-led incentives for non-incremental technical change. *Research Policy*, 38, 700–709.
Perlin, J. 1999. *From Space to Earth: The Story of Solar Electricity*, Ann Arbor, MI, AATEC Publications.
Shimada, K. 1983. *Photovoltaic Research and Development in Japan*. National Technical Information Service, U.S. Dept. of Commerce.
Shimamoto, M. 2014a. Chapter 3: Case study –managing technology development. *Keikaku no sohatsu: Sunshine keikaku to taiyoko hatsuden (National project management: the development of photovoltaic power generation system under the Sunshine Project)*. Tokyo: Yuhikaku (in Japanese).

Shimamoto, M. 2014b. Chapter 7: The politics of creating new significance. *Keikaku no sohatsu: Sunshine keikaku to taiyoko hatsuden (National project management: the development of photovoltaic power generation system under the Sunshine Project)*. Tokyo: Yuhikaku (in Japanese).

Shum, K. L. & Watanabe, C. 2007. Photovoltaic deployment strategy in Japan and the USA—an institutional appraisal. *Energy Policy*, 35, 1186–1195.

Sugiyama, T. 2018. Decreasing cost of PV: Who should take credit?: The role of government in technological progress. International Environment and Economy Institute.

Suzuki, J. & Kodama, F. 2004. Technological diversity of persistent innovators in Japan – Two case studies of large Japanese firms. *Research Policy*, 33, 531–549.

Takahashi, K. 1989. Sunshine project in Japan-solar photovoltaic program. *Solar Cells*, 26, 87–96.

Tatsuta, M. 1996. New sunshine project and new trend of PV R&D program in Japan. *Renewable Energy*, 8, 40–43.

Watanabe, C., Wakabayashi, K. & Miyazawa, T. 2000. Industrial dynamism and the creation of a "virtuous cycle" between R&D, market growth and price reduction – the case of photovoltaic power generation (PV) development in Japan. *Technovation*, 20, 299–312.

Watanabe, C., Kishioka, M. & Nagamatsu, A. 2004. Effect and limit of the government role in spurring technology spillover – a case of R&D consortia by the Japanese government. *Technovation*, 24, 403–420.

6
GERMAN DEMAND-PULL

Germany contributed to the development of inexpensive PV by creating a far bigger market opportunity than had ever existed. After two decades of developing new organizations and building an advocacy coalition among them and others, a policy window opened when the Green Party became a ruling partner in 1998 leading the German Parliament to pass the Renewable Energy Law (Erneuerbare-Energien-Gesetz, or EEG in German) in March of 2000. The renewable strategy has its roots in the 1968 student protests, the anti-nuclear movement of the 1970s, and the emerging environmental movement catalyzed by the Chernobyl accident in 1986. Support for PV was originally grassroots and local, but was later institutionalized at the federal level and reached its apex in the revisions to the EEG in 2004, which favored PV. This long-term grassroots movement underpinned a rising expectation that the government would support renewables and PV, in particular. The direction of policymaking was clear even in the late 1990s when firms like Q-Cells were founded. From 2004 to 2012, the EEG supported the adoption of over 30 gigawatts of PV in Germany with a subsidy program that totaled over 200 billion euros. The EEG transformed the world PV market, accounting for half of global PV installations in the peak period of the program, 2004 to 2010, during which the world market grew by a factor of thirty.

The EEG catalyzed a global process of learning by doing and created opportunities for massive economies of scale. For the first time, equipment providers could design machines specifically for PV applications rather than repurposing them from the semi-conductor industry. This allowed PV producers to automate their production processes. The size of the new market opportunity interested Wall Street investors and venture capitalists. It enabled PV startups to hold initial public offerings, raise hundreds of millions of dollars, and invest in scaling up production to levels that were orders of magnitude larger than had been seen before. All of this reduced the costs of PV, which in turn expanded demand as lower prices attracted new adopters, which enabled further scale. From the EEG's inception in 2000 to its decline in 2012, prices of PV modules dropped to 16% of their pre-EEG level. In retrospect Germans positioned this as a benefit to the rest of the world.

> Indeed, the German people are paying significant money, but in Germany, we can afford this—we are a rich country. It's a gift to the world. – Markus Steigenberger, Agora Energiewende (Gillis, 2014)

German PV producers surpassed Japanese producers and became the largest in the world. Solarworld, Q-Cells, and others eventually ran into trouble in their limited ability to lower costs as much as their new competition in China did. Despite substantial advantages associated with being located in the most important market in the world, Q-Cells—once the largest—was never able to scale up to the level of its Chinese competition. It had locked in expensive silicon procurement contracts, diversified excessively, was late to target ambitious cost reductions, and, most importantly, was eventually cut off from capital during the global financial crisis, just as its Chinese competitors were investing tens of billions in scale up.

Still, in the longer-term, domestic beneficiaries of the EEG thrived, particularly the local solar installers who were overwhelmed with demand. The other sector that took full advantage of the opportunity was PV equipment providers such as Centrotherm, Roth & Rau, Rena, and Schmid. They also benefited from the talent emerging from Germany's education system and its steady investment in PV R&D since 1981. The machine makers contributed most to industrializing PV manufacturing by designing PV-specific machinery, which they sold to German producers and later to the Chinese. In this period the technological frontier moved away from cell design and toward equipment design. Around 2004, these firms began selling "turnkey" systems, which integrated several machines into a production line and required less know-how by producers. The rapid increase in demand that the EEG caused required speed on the part of producers, so the turnkey option sold well and enabled new companies to enter the market.

More broadly, the EEG showed the world that PV was a serious technology that could be scaled up, developed as an industrial production process, and integrated into electric grids in large amounts. The policy itself, particularly the feed-in tariff component, was used as a model that could be implemented in a wide variety of countries (Steinbacher and Pahle, 2016)—at first in Spain and Italy and later in Asia and Africa. Some argue that the success of the EEG showed that transforming the energy system was easier and less expensive than anticipated (Hager and Stefes, 2016). In this interpretation, the EEG enabled the Energy Transition policy adopted by Germany in 2010 (Morris and Jungjohann, 2016) as well as the Paris Agreement in December 2015 (UNFCCC, 2015). Because this wide array of beneficial impacts has had their biggest impact outside Germany, the EEG has been called Germany's "Gift to the World." One interviewee said it may turn out to have been the most momentous technological development in Germany since the printing press.

From a national innovation systems perspective, Germany used its long history of education, research, and engineering prowess to develop a production equipment manufacturing industry and design of production process lines that continues to lead the world, a decade after the largest producers in Germany lost their lead. Germany was remarkably open to new policy ideas, borrowing important aspects of the design of the EEG from California (guaranteed purchase prices) and Japan (declining subsidy levels). The development of the EEG was more of a bottom up process than policymaking in any of the other PV countries. It evolved over 25 years beginning with protests, moving to local initiatives, and eventually national policy—a process that overcame the objections of large electric utilities. The costs of the EEG were substantial and borne by rate payers. Another hallmark of the German innovation system was a long time horizon; the EEG was reinforced by the policy of making an energy transition (Energiewende, EW) with targets for renewables, efficiency, and CO_2 emissions over forty years. It also was savvy on political economy considerations, leaving large electricity consumers out of the costs burden. Finally, the EEG and EW reflect a cultural ability to manage transformative change that has roots in the post-war rebuilding after WWI, after WWII, the rise of the EU, and the integration of the former East after the fall of the Berlin Wall.

Bottom-up: 68ers until the late-1980s

Public and political support for renewable energy in Germany emerged out of the "68er" student movement and the anti-nuclear movement that followed in the 1970s and 1980s. Grassroots organizing became a highly successful model and thus, much of the early political support for renewables arose at the municipal and Länder (provincial) levels.

The 68er movement was a student protest movement in 1968 at West German universities that echoed aspects of the protest movements in France and the US opposing the Vietnam war. It was a peaceful protest against perceived authoritarianism and employed grassroots organizing as a challenge to traditional German corporatist politics. Hans Josef Fell, one of the two architects of Germany's renewable energy law, describes it as a "culture of revolution" (Davidson, 2012) and the other architect, Hermann Scheer says:

> The jointly shared experience was that even in countries with democratic institutions, governments were not willing to do what had to be expected of them. Without this background, it is not possible to explain the process that led in Germany to a popular movement for renewable energies that helped create the democratic basis for legislation to mobilize PV. (Palz, 2010).

The 68er movement had a larger impact in Germany than in other countries, as its effects persisted for decades. For renewable energy it raised the possibility of grassroots protest and organizing as alternatives to traditional politics. The small scale of solar fit this new model.

A grassroots anti-nuclear movement arose in the mid-1970s in Baden-Wurttemberg and centered in Freiburg. The approach was similar to that of the 68ers in that it employed grassroots organizing and peaceful protest. Just as the 68ers were connected to geopolitics in its anti-Vietnam sentiment, the anti-nuclear power movement was closely linked to the Cold War. Germany was on the front lines and if tactical nuclear weapons were used, Germany would be affected first. The focusing event for this movement was the occupation of a construction site, which successfully defeated plans to build a nuclear reactor at Whyl in February of 1975. The mantra in the local dialect "Nai Haemmer Gsait" (we said no) became a rallying cry and similar organizing tactics were used in protests thereafter. It became a bottom-up movement that emphasized moral conviction, small producers, and large public support despite substantial costs (Hager and Stefes, 2016). That electricity decisions in Germany were still guided by the Nazi-era 1935 German Energy Industry Act stimulated the anti-authoritarian aspects of the movement.

The oil crises in 1973 and 1979 added a further element to the movement in that they made clear the need to diversify away from oil. Because the protests essentially took nuclear off the table, because coal contributed to acid rain, and because natural gas was imported, renewables offered a positive direction around which the movement coalesced. The protest movement was thus atypical NIMBYism because it proposed an alternative, was technologically oriented, and forward looking. This pro-technology direction led to the establishment of institutions, such as the Fraunhofer Institute for Solar Energy Systems (Fraunhofer-ISE) in Freiburg, founded by Adolf Goetzberger in 1981. Goetzberger had returned to Germany after working on transistors at Bell Labs. Fraunhofer-ISE soon became a member of the Leibniz Association, the official group of German research institutes that are eligible for federal research funding. Even earlier, in 1975, the "solar energy working group" was founded, consisting of five companies: AEG, Dornier, Philips, RWE, and BBC, and later Bosch and Wacker.

The Chernobyl nuclear accident in April 1986 further heightened opposition to nuclear power and by then a direct connection to supporting renewables was firmly in place. Despite the accident occurring over one thousand miles away, its impacts were felt locally. A monitoring site atop a mountain near Schoenau was one of the first sites to detect the abnormal levels of radiation. Large parts of Germany were exposed to contaminated rain and local mushrooms and boar meat were avoided for years. This led to the establishment of "Parents against nuclear power." By the end of 1986, the German federal government established the Environment Ministry. Chernobyl was a milestone event for environmentalism in Germany and it gave public and political support for renewables a major push that would last for two decades. When the Fraunhofer-ISE faced closure in the late 1980s, the experience of Chernobyl kept it alive.

Chernobyl was the focusing event, but there were other drivers that made the environmental movement in Germany in the 1980s more than a fringe element in German politics. In the early 1970s, The Limits to Growth studies by the Club of Rome influenced thinking about tradeoffs between economic growth and other social objectives (Meadows et al., 1972). The die off of trees in the Black Forest in the early 1980s from acid rain exposed the downsides of burning coal. That this event occurred in Baden-Wurttemberg further concentrated the local engagement. In November 1986, just six months after Chernobyl, a fire at a chemical plant near Basel, Switzerland washed 1,000 tons of pesticides and other chemicals into the Rhine turning the river red as it entered Germany in that same province.

This array of factors began to generate institutional responses. In 1987, the German Physical Society released its first report on climate change (DPG, 1987) leading Chancellor Kohl to call climate change the world's most important environmental problem. Climate change has remained prominent on the political agenda in Germany ever since. The link between climate and renewables was made early on, for example by the Enquette Commission on climate in the late 1980s. The drivers in the 1970s led to the founding of the Green Party in Karlsruhe in 1980, which merged with the East German citizen movement, Bündnis 90, in 1990. The Green Party has received 5–10% of the parliamentary vote in almost every election ever since, big enough to potentially be a member of a coalition government.

Environmental non-governmental organizations (NGOs) also began to emerge. Greenpeace was active in Germany and in line with the grassroots origins of German environmentalism, solar NGOs emerged early on. A traditional trade organization, the German Solar Energy Industries Association, was founded in 1978. The Öko Institute was founded in Freiburg and had a focus on promoting renewables with an explicit aim to counter governments and utilities who were thought to oppose renewables. The Förderverein Solarenergie was founded in 1986 and developed and promoted the idea of using government payments to cover the cost of renewables, a notion that was later included in feed-in laws. In 1988, Herman Scheer founded Eurosolar to lobby politicians for policies supportive of renewables. Some of this advocacy was justified under the notion of "Ordo-liberalism," that the state needs to intervene to make markets fair (Morris and Jungjohann, 2016).

These drivers and the institutions they engendered steadily built support for an emerging policy regime supportive of renewables. Solar policy in Germany in the 1970s and 1980s almost entirely consisted of energy R&D funded by the Federal Ministry for Research and Technology, which began in a major way in 1974, peaked in 1982, and subsequently declined as the conservative Kohl administration began its 16-year rule. R&D was strong in the US in this period but a substantial amount was invested in Germany as well. A broad competence in solar technology emerged in research institutes and companies despite most of the R&D being spent on nuclear and other technologies. A little bit of solar R&D went a long way. Other than R&D, renewables had almost no support in the 1970s and 1980s, certainly not at the federal level (Jacobsson and Lauber, 2006). The electricity sector, dominated by coal and nuclear, was generally hostile to renewables throughout the 1970s and 1980s. Some activities beyond R&D began in 1983 with the first large-scale (300 kW) PV demonstration plant. Subsequently this program funded 70 large PV installations that were built between 1986 and 1995.

Beyond R&D, some top-down support for renewables began to develop during this period. A 1980 Öko Institute paper was the first to describe an energy transition (Krause, 1980). Their idea was to phase out oil and nuclear power by switching to the soft energy paths advocated by Amory Lovins (Lovins, 1976). A key, and distinctly realistic, aspect was that this would involve a slow transition over decades. In addition, an Enquette Commission in 1980 concluded that energy efficiency and renewables should be the first priority, even while maintaining the nuclear fleet. A five-year study from 1981 to 1986 confirmed the prioritization of efficiency and renewables but changed to advocating dropping the nuclear fleet. That study was released just prior to the Chernobyl accident. In 1994, an environmental protection clause was added to the German constitution.

This period also saw nascent private sector activity. Early German PV companies Siemens and ASE (Applied Solar Energy: Angewandte Solarenergie) acquired US PV firms that had state-of-the-art technology. Alpha Real, a PV company in German-speaking Switzerland, launched "Project Megawatt," a program to install 333 systems on Swiss rooftops. Founder Marcus Real had visited the US in 1986 to see the Rancho Seco PV project near Sacramento. Alpha Real became one of the first corporate advocates for rooftop solar and the first to push the idea of what later became known as "net metering."

Early rooftop programs: 1988–1999

More importantly, from 1990 to 2000 cities led the way on PV adoption in Germany. Aachen was the pioneer with the first community-based introduction of PV. In 1992, Wolf van Fabeck proposed a 2 Deutsch Mark (DM), about~$1.3, per kWh tariff, about 20 times the retail rate. Von Fabeck and other citizens mobilized their city council to pass several resolutions to require that the local utility company provide reasonable rates that covered the full cost of an investment with a standard profit added on for solar power. The Aachen City Council adopted the plan in 1994. Aachen was not the first city to achieve full-cost compensation for solar but was used as a model for dozens of other municipalities who followed its lead. PV activists had learned how to organize on a local level, they learned about the technology and began to understand the economics—of PV and of the markets in which it would compete. Activists enabled the local governments to write a law with strong support from Hermann Scheer. Aachen served as a model for municipalities across Germany to adopt PV policies that subsidized the cost of installing systems. For example, in 1993 Hans-Josef Fell applied the Aachen model to his hometown of Hammelburg, Bayern using a compensation scheme of 1.89 DM per kWh ($1.15 per kWh) for 20 years, which was about twice the level that the EEG passed a decade later. One consequence of the utility programs is that they stimulated firms to enter the PV industry, as module manufacturers and installers. A second consequence is that the diffusion of the Aachen model to other cities, and the PV adoption it stimulated, legitimized PV and made clear that there was substantial public interest in PV.

Nearly simultaneously, the federal government began experimenting with applying a similar scheme nationally. Like the 1000 Roofs program, they borrowed heavily from PURPA and the California Standard Offer Contracts, particularly ISO4, which had been very effective in stimulating private sector investment in wind power (Chapter 4). Like the California contracts, it would be technology neutral, not just for PV. Since PV and other renewable technologies were primarily considered R&D projects by the federal government, despite installation programs by municipalities, the policy was intended not as a deployment program but as a "mass demonstration" project. Its portrayal as an extension of renewables R&D helped deflect stern criticism from the Economics Ministry, which was skeptical of subsidizing specific technologies and concerned about the government picking winners.

Like the ISO4, the core concept was a feed-in law, which would require utilities to allow generators of renewables to connect to the grid and buy electricity from them at a specified rate. The rate was set at 90% of the retail electricity rate, even though the scheme was providing payments in wholesale markets where prices were a factor of two to three lower. The rate was guaranteed for 20 years. Interconnection agreements were standardized to ease the process for small generators. The justification for this policy is that it would level the playing field for small generators. Conservatives were brought on board with the argument that it would help account for the pollution externalities associated with fossil generation. Despite the generous prices offered for electricity, PV did not benefit from this scheme, and almost all of the deployment stimulated was wind power.

The Electricity Feed-in Law of 1990, the Stromeinspeisungsgesetz (StrEG) went into effect on January 1, 1991. It was the world's first renewable electricity feed-in tariff (FiT), even if the similarly designed ISO4 came first, it was not a FiT in name. As in other schemes, once a large amount of wind power began to be deployed, utilities attacked the law both through the courts and through the legislative process (Jacobsson and Lauber, 2006). Preussen Elektra challenged the law under EU anti-subsidy rules and was ultimately unsuccessful in 2001 when the European Court of Justice ruled that the law did not constitute state aid. On the other side, a 1997 proposal to reduce the tariffs revealed that an advocacy coalition had begun to develop when large demonstrations ensued involving community groups, metal workers, as well as renewable energy industry groups and investors. The awareness that renewables had broad public support allowed plans for more ambitious policies to proceed. After lobbying policymakers for support, ASE/RWE-Schott Solar built a new 20 MW cell plant in 1998 and that same year, Shell built a 10 MW plant in Gelsenkirchen (Jacobsson et al., 2004).

With multiple municipalities devising their own schemes, the over-subscription of the 1000 Roofs program, and the innovative design of the StrEG, the emerging advocacy coalition began to lobby for something bigger. Companies in the renewables industry threatened to leave Germany if a new subsidy program was not started. Greenpeace and Förderverein Solarenergie pushed for a new program. In 1993, through his PV lobby organization Eurosolar, Hermann Scheer proposed a 100,000 roofs scheme, but it was rejected.

Something much bigger was to come and that hinged on the outcome of the September 1998 elections in which a coalition government emerged comprising the center-left SPD and the Green Party, a coalition that lasted until 2005. After more than twenty years of grassroots organizing, the Greens had a formal hand in national politics. The agreement establishing the coalition included support for renewables, a 100,000-roof program for PV, and a phase out of nuclear power, all of which came to fruition within a few years (Staiss and Räuber, 2003). The 100,000 Roofs Program was far less effective in stimulating adoption than other policy instruments in that only 3,500 loans were made by the end of 1999 and the scheme was cancelled in 2003. Still, companies began to notice that political support for their industry was now substantially larger than it had been.

EEG: 1998–2012

Hermann Scheer, a Social Democrat, was a member of the German Bundestag from 1980 until he died in 2010. Looking back on the success of the EEG, Scheer argues that "what is important is social acceptance and vision" and spoke of the need for "emancipatory motivation" (Palz, 2010). He was a student leader at the University of Heidelberg, one of the hotspots of the 68er student movement. He made a direct connection to the activism and grassroots organizing of that movement, talked about PV as emancipation, and referred to the moral philosophy of Kant. *Solar Economy*, one of his books, anticipates much of what later happened (Scheer, 2004).

Fell was younger, only a high school student in 1968, but followed the 68er student movement in universities. He was a member of the Green Party from 1998–2013. He felt close to its goals and was also influenced by the Limits-to-Growth studies and by 1973 oil crisis. He admired President Carter for the prominent role he made for solar in the late 1970s. Seeing the US's solar program vindictively taken down under Reagan made Fell realize that powerful interests would oppose the solar vision despite its broad appeal. As a result, he structured the EEG to include specific price and other quantitative parameters in the law, rather than delegated to a ministry or regulatory agency, to avoid risk of interference by opponents, particularly the Economics Ministry. That insight was crucial for getting the EEG adopted, maintaining its stringency, and for keeping it in place for well over a decade.

In designing the EEG (RESA, 2001), Scheer and Fell drew on two decades of prior policies, in particular the previous municipal German subsidy programs, the Japanese rooftop program, and the US ISO4 contracts. Like those contracts, the EEG required grid operators to purchase electricity from small producers. Like the municipal feed-in tariffs, it guaranteed the rate that grid operators would pay to PV electricity producers. Learning from prior programs, it distinguished that rate by technology—higher for less mature technologies like PV and lower for more mature like wind. Rooftop PV systems (up to 5 MW) would receive 0.99 DM per kWh ($0.50) in 2001, which was well over twice the retail electricity rate. To limit risk to investors, that rate was guaranteed for twenty years. To appease opponents of the law—including conservatives, the Economics Ministry, and the Competition Directorate at the European Commission—the law borrowed a Japanese policy component and stated that the guaranteed rates would decline by 5% each year. The policy was also constrained by including a sunset provision that would end the FiT once PV capacity reached 350 MW. The above market rates for solar electricity were funded by a surcharge on rate payers, which over time has worked out to about 20 euros per household per month (Unnerstal, 2017). In another political compromise, large electricity intensive manufacturers were exempted from the surcharge under the argument that they would be unfairly disadvantaged against foreign competition.

In getting the EEG adopted, Scheer and Fell convinced their fellow members of the Social Democrats and the Green Party that the policy would allow PV to become a serious industry. One angle was to accentuate that the policy would provide a rate of return to investors in PV of 12–15%. That anyone could make such a good return, even small homeowners, appealed to a broad set of members of the German Parliament. They also made the case that opportunities would exist in the former East of the country where additional subsidies were available for establishing manufacturing. Some conservatives were also convinced, in part because of the high solar resource in the more conservative southern parts of Germany. Other conservatives, as well as electric utility interests, dismissed PV as too small to matter, so they didn't seriously oppose it. The EEG went into force on April 1, 2000. Adoption and subsequent implementation of the EEG catalyzed opposition to it. The Federation of German Industry, the association of electric utilities, and the Economics Ministry (run by the SPD) criticized the EEG. Yet the Greens improved their results in the 2002 elections while the SPD's declined, leading the Greens to take the lead on renewables policy moving that to the Environment Ministry (Bundesumweltministerium) rather than the Economics Ministry. Still the utilities especially worked hard to reduce the impact of the law for the next ten years.

The rising strength of the Green Party allowed them to push for a more stringent version of the EEG. The Renewable Sources Act Amendment passed in July 2004 maintained the structure of the EEG, increased the feed-in rates for PV, included more differentiation in rates across technologies, and expanded the number of entities who were excused from paying for the EEG. In addition, a new renewable energy target of 20% by 2020 was introduced. These revisions sparked the EEG's "gold rush moment." Installations grew by a factor of four from 2003 to 2004 and transformed the industry. For example, in 2002–03 Ersol had been seeking customers who would sign on to long-term contracts. In 2004–05, demand became so large that they were happy to cancel contracts they had signed in order to meet demand for new larger ones. The 350 MW target was reached within three years of the 2000 law and was increased to 1000 MW in 2002. Renewables went from 6% of electricity to 27% in 2014. Financing for these systems came from local banks, Landesbanken (state banks), and low interest loans by Germany's development bank, KfW (Quitzow, 2015).

Some argue that demand was too much too fast. The new tariff of $0.54 per kWh in 2004 was perhaps too high. One interpretation is that it took so long to get the EEG and revisions passed that the rates quickly became outdated as PV prices fell and demand became so large. Choosing the right feed-in rate was being done by politicians not by markets. This required consensus and compromise amidst changing parliamentary majorities and coalitions. Markets, and particularly manufacturing costs, were moving much faster than politicians. One outcome was entry into the PV industry as the industry matured. Cell manufacturers proliferated. Manufacturing investment concentrated in "Solar Valley" near Thalheim in Germany's former East, in Freiberg, near Dresden, and in Erfurt. Some complained that this entry involved "C and D-class players" who saw an easy 12–15% return. Another criticism is that the rapid growth enabled the rise of Chinese producers who were able to scale much faster than the Germans—mostly because, especially in 2009–10, Chinese state banks provided local companies with more capital at lower interest rates than the financial sector provided German companies access to capital. Utilities seem to have completely underestimated the demand for solar. "They missed it" and never considered PV a serious technology. A prevailing opinion was that PV was interesting but would take 50 years to be important.

In the longer term, the EEG enabled the development of more stringent renewable energy targets and more broadly, of the Energiewende. With its roots in the concept of "soft energy paths" (Lovins, 1976), originally coined as a term in 1980 (Krause, 1980) and published by the government as a "Concept Paper" in 2010, the Energiewende (EW) is Germany's comprehensive 40-year plan to increase renewables, phase out nuclear power, and reduce carbon dioxide emissions (Morris and Jungjohann, 2016). Under the EW, renewables goals were 35% by 2020, 50% by 2030, 65% by 2040, and 80% of electricity by 2050. In 2018 a coalition government agreement led these goals to be upgraded to 65% by 2030. These targets would have been unthinkable 20 years earlier. The experience with the EEG made them possible. Along with energy efficiency, the renewables component of the EW is the target most on track to being met (Unnerstal, 2017). Costs remain an issue and CO_2 emissions have not been falling enough, at least in the near term due to nuclear shutdowns and renewables displacing natural gas rather than coal.

Internationally, the EEG has enabled Germany to export its policy leadership (Steinbacher and Pahle, 2016). In the early 2000s renewables were still quite obscure outside Germany. The International Energy Agency (IEA) modeled energy systems and assessed cost reductions. Its publication *Energy Technology Perspectives* documented the progress in renewables (IEA, 2006). At the IEA, Clas-Otto Wene was a strong proponent of the learning curve for PV and published an influential book (Wene, 2000). Still renewables did not feature in IEA's most high-profile work, the annual World Energy Outlook, and continue to be downplayed today (Creutzig et al., 2017). The EEG began to legitimize PV by showing what could happen on the ground in a major way, not just in niche markets or in models. Partially in response to the IEA's obstinance, Germany was behind a new international organization, the International Renewable Energy Agency (IRENA) with the goal of 700GW of renewables by 2050. The main role of IRENA was to convince governments that there could be a significant role for renewables. Another step toward legitimacy was the Intergovernmental Panel on Climate Change, which published a special report on renewable energy in 2010 (IPCC, 2011). The experience of the EEG in Germany made all these international developments increasingly credible since they showed that there was demand for solar, that costs could come down, and that policy could catalyze the interaction of both.

Together the 2000 EEG and 2004 revisions were a game changer for the PV industry, its effects extending far beyond Germany as we will see in Chapter 7 on China. But why did it happen then?

A main explanation is that in 1998 a policy window opened. The first small-scale PV and wind systems had become available and since wind costs had been falling, PV seemed ready for wide adoption. PV advocates anticipated that a learning curve would allow for further cost reductions. PV technology had become well known; people had used calculators, it worked in satellites, and systems were still operational after over a decade of continuous use. The first PV system in Germany installed at Pellborn on the North Sea in 1980s was still operating in 2000. Further, Scheer and Fell were confident that a federal program would work because they had seen it work on the local level. Experience with these early projects made clear to them that utilities would not support renewables and that citizens would need to participate directly. On the other hand, Fell knew that over a hundred pro-solar NGOs had formed in his home state of Bavaria alone, showing that the public appetite for PV was strong.

Given that the first rooftop subsidy program was in 1988, a related question then is why did it take so long to get to the passage of the EEG in 2000? People needed to gain experience with PV, to see that it worked, that systems did not degrade, that PV could work with the electric grid. The industry had to mature. For example, a diverse landscape of two dozen manufacturers emerged in the wind industry. It had to initiate economies of scale to show that future cost reductions were likely. Investors needed time to sort out their own expectations about the technology, future costs, and the markets in which it would compete. People had to become comfortable with the idea that the market would decide which technologies (e.g. wind vs. solar) would be adopted, rather than the state choosing specific demonstration projects. In a sense, the EEG and the PV boom that ensued appeared to have happened overnight, but there was a long evolution behind it. Still, there was not a prominent public debate about it before launch. By 2000, it was clear that PV involved no "technology risk" and its success was all just a matter of cost.

Eventually though, the EEG faced backlash. A victim of its success, the EEG became a high-profile target and its ramp down began soon after the 2004 revisions. A series of developments led to increasing opposition to the EEG. The September 2005 elections led to a new governing coalition of the two largest parties, the center-right Christian Democrats (CDU) and the center-left SPD. Both parties had strong ties to the coal industry. The Greens were out. In part due to rapid demand for PV, the cost of purified silicon rose dramatically from 2004 to 2008, raising input prices and leading to PV module prices that were steady instead of continuously decreasing as advocates had promised. The learning curve, one of the main justifications for the EEG, appeared to have reached its end. Utilities started to gain more sway in renewables policy debates. They had missed the importance of PV in the formative years of the EEG, especially from 1998 to 2000, but by 2007–08 they had turned aggressive in stopping it. By 2010 multiple parties were making strong public statements against renewables.

Discussion of the costs of the EEG and of the EW became more widespread. The EEG surcharge—the payments from residential electricity consumers to fund the PV and wind subsidies—had been growing: from 0.54c per kWh to 1.32c per kWh. A recent comprehensive estimate found that the EW would cost about 0.25% of GDP from 2015 to 2050, about half of which was due to the nuclear phase out (Unnerstal, 2017). In comparison, German reunification was about 1.2% and coal subsidies about 0.3%. Households would pay about 20 euros per month. In addition, there were contentious discussions about the distributional and regressive aspects of those costs—the costs of the subsidy constituted a high share of income for the poor and little for the wealthy who were far more likely to adopt solar. But the overall societal costs were barely discussed before 2010.

In the mid-2000s, PV supporters had argued for dynamic adjustment to subsidy levels to avoid excessive deployment volumes in individual years. There was indeed a delicate balance to be struck between maintaining a strong policy instrument and avoiding high investment volumes that might create bottlenecks and thus raise prices. The first revisions to the EEG that substantially reduced its stringency began in 2009 with the Renewable Energy Sources Act. The "degression" rate for PV—the annual reduction in feed-in levels—was changed from 5% to 8–10% and the rates became flexible, in that they could be adjusted further downward if the amount of new PV adoption exceeded "corridor" deployment targets. The tariff levels were cut further in July 2010 by 11–16% with the PV Act of 2010 (Figure 6.1). The PV Interim Act of July 2011 cut the tariff further and doubled the charge to electricity consumers. The Renewable Energy Sources Act of 2012, under a new all-right coalition of the CDU and Free Democrats, aimed to prepare PV for unsubsidized markets and introduced a market premium scheme—similar to the US model of the Production Tax Credit. The act cut the tariff further and for the first time allowed grid operators to curtail PV if the grid became congested. The 2013 PV Act cut tariffs by 30% and introduced a hard cap on PV of 52GW. The 2014 Renewable Energy Sources Act initiated EEG 2.0 with a shift to auctions and pre-specified deployment corridors.

By 2013, those who opposed PV in Germany could declare victory. The German PV market after growing by more than 50% per year from 2004, stalled in 2011, growing at only 1% and then 2% in 2012. This sudden end to rapid growth

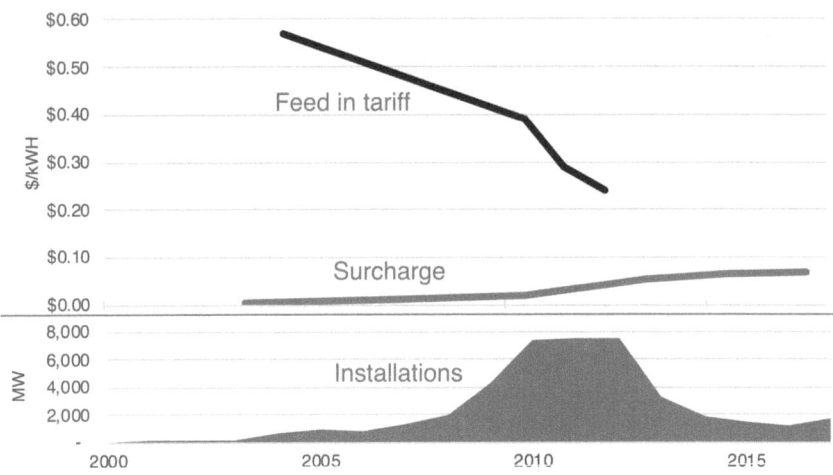

FIGURE 6.1 Subsidy levels of the German feed-in tariff (FiT), the surcharge applied to rate payers, and installations in Germany.

shocked PV producers but particularly German ones because they were more exposed to their domestic market. Q-Cells, Sunways, Schott Solar, all folded within a year. Starting in 2013, the German PV market got smaller each year and continued to shrink until 2017. It was not a surprise that after a few years of stagnation from 2006 to 2009, the learning curve went back to work thereafter and led to rapid costs reductions from 2010 to 2013. By then, however, China had become the world's largest producer.

Q-Cells: a star but why couldn't they scale up further?

Q-Cells was the brightest star of the EEG. Although it preceded the EEG, it also lived and died by it. Like Scheer, Reiner Lemoine, a founder of Q-Cells, was a 68er. He graduated from the Technical University of Berlin in aerospace engineering and became involved in Wuseltronik, an R&D lab in Kreuzberg, also known as a "socialist engineers collective." In 1996 he founded Solon AG, which produced modules from cells that he and colleagues bought. Soon Lemoine and two others at Solon, Paul Grunow and Holger Feist, started planning a new company, where they would produce their own cells with a focus on high "quality." They named these high quality cells Q-Cells.

After failing to raise finance for a production facility in Berlin, in November 1999 they founded Q-Cells in Thalheim, outside Bitterfeld, an area that was later to become known as "Solar Valley." The Mayor of Bitterfeld, Manfred Kressin, was enthusiastic and facilitated financing and identified a location with vast room for expansion. They raised 60,000 euros in startup funding and obtained seed funding from angel investor Immo Stoher who had invested in Solon and owned 20%. They hired as CEO Anton Milner, a chemical engineer and ex-McKinney consultant who worked in the oil industry. Since they were focused on making cells, their initial customers were PV module makers. Q-Cells was a sexy start up in a world of PV companies as subsidiaries of global oil companies. They were full of young people, moved fast, and threw lavish parties at solar trade shows.

They also worked quickly and by July 2001 were operating a 10 MW line. They secured a second round of funding in 2002 from Good Energies, a philanthro-capitalistic fund founded by heir to the C&A department store fortune. Good Energies would own 49% of Q-Cells by 2006. Q-Cells doubled capacity in 2002 and won a start-up award, for which part of award was pro bono consulting from McKinsey. With Milner as CEO, this link paved the way for an influx of ex-McKinsey consultants to the firm, which some employees would later criticize as leading to a short-term focus.

Nevertheless, Q-Cells expanded production capacity ten-fold from 2002 to 2005 to nearly 300 MW over 10% of the global total. They went public in 2005, completed their fourth production line in 2006, and sold a billion dollars of cells in 2007. The established solar players, BP Solar, Shell Solar, RWE-Schott, were slow and conservative and did not take the upstarts seriously. Q-Cells, along with other pure play PV startups like Solarworld and ErSol were aggressively focused on growth. The established players started to go out of business.

Getting big fast continued to be a successful strategy from 2005 to 2008. During the EEG boom, renewables were adding 1% of German electricity supply per year. Headhunters were calling PhD students to drop their studies and move to firms like Solarworld and Q-Cells. Engineering expertise allowed Q-Cells to overcome any barriers to scaling up production quickly. In any case, scale up was not technically challenging. For making wafers and cells, adding machines and moving to bigger wafer size worked well. For those producing silicon ingots, scale up involved adding furnaces and pullers and occasionally moving to larger ones. Further, plenty of knowledge about PV production was in the public domain. Universities in the US, Germany, and Australia trained students who went all over the world. Knowledge was also transmitted by equipment manufacturers. Finally, as seen in Chapter 4, one should not discount the idealism that pervaded many of those in the industry, which encouraged openness.

Q-Cells was the largest cell producer in the world in 2008. But along the way to becoming a giant, Q-Cells made two critical decisions that affected their long-term future. Both were responses to a major disruption in the industry, the rising cost of purified silicon. These decisions not only increased input costs, but also affected the willingness of German cell and module producers to expand production. As Germany's largest cell producer, Q-Cells, stated in its 2006 annual report (Quitzow, 2015),

> Our policy is to only expand our production capacity when we have access to the corresponding raw materials.

First, they entered into long-term supply agreements for poly-silicon. In mid-2005, they signed supply agreements for 900 tons of silicon, equivalent to 740 MW of cells, or about three years of production. That amount would have cost $100 million at spot prices. In a 2006 SEC filing soon after their IPO, Q-Cells reported,

> We are assuming that bottle-necks in silicon availability are also likely to occur in 2007 and 2008. Compared with our competitors, we are well-placed in covering our supplies of silicon and wafers because of our positioning and our excellent, close collaboration with our main suppliers dating back over many years to continue achieving high growth rates and production volume.

The following year, they signed an agreement with Elkem Solar in Norway for over 2,000 tons annually from 2008 to 2018, which was 10 GW worth of cells, about 10 years of production at their anticipated ramp up rates.

> We are convinced that the long-term contract announced at the beginning of February will give Q-Cells a considerable advantage.

They also signed on for 7,400 tons from REC in Norway, 6,000 tons per year from BSI in Canada, and a 10-year contract for more than 6 GW of wafers from LDK in China. Q-Cells obtained fixed prices for the 2008 and 2009 deliveries while 2010–11 prices would be an average of a pre-agreed contract price and the spot prices at delivery. Given the crash in silicon prices that began in late 2008, these contracts, totaling around $2 billion, would turn out to be very expensive for Q-Cells. Q-Cells financed these purchases with a 500m euro bond issue, some of which it also used for R&D.

Second, Q-Cells diversified into a broad array of alternatives to crystallized solar. Thin film was one focus, Cadmium Telluride (CdTe) and copper indium gallium selenide (CIGS) particularly. In 2005, they invested $6 million in CSG Solar and built a 25 MW thin film line. In 2007 their subsidiary Sonto GmbH expanded its micro-morph thin film line to 60 MW. They expanded a ribbon silicon plant in Bitterfeld with their US partner Evergreen Solar. They produced CIGS cells with their subsidiary Solibro and built a new line for another thin film subsidiary Calyxo in 2008. In 2008, at the most rapid period of market growth, Q-Cells invested in R&D for next generation PV,

> In addition to investment in the Company's core business, the funds will be spent on constructing a research and development line and on commercializing new technologies.

While a focus on the longer term seems prudent, it is striking that Q-Cells would invest so much in alternative PV technology when competitors in China were laser focused on increasing production capacities for crystalline silicon PV. One interviewee said they did too much "dabbling" in different technologies. Another said that Q-Cells did not believe that crystallized silicon PV prices could ever drop as much as they did. Chinese firms did not share Q-Cells' pessimism about the potential cost-reduction in crystallized PV and turned out to be right and, as a result, Q-Cells' shift to alternatives put the company's competitive position in jeopardy.

Q-Cells began to suffer from these decisions in the first quarter of 2009, when it reported its first loss, which grew to a full year loss of $1.9 billion. It laid off 500 employees and cut working hours to 80% for those it retained. CEO Anton Milner resigned in March of 2010. Prices were falling rapidly as its competitors were able to cut costs much faster than Q-Cells could. Q-Cells had negative gross margins in 2009 as a result of production costs that were 15% higher than competitors and the silicon procurement contracts to which they were bound. This meant that Q-Cells was losing money on every additional cell it sold. Milner had emphasized the "Q" in Q-Cells and marketed that they were German made and, thus, more reliable, but the lower costs won. Under their initiative, "Q-Cells Reloaded," they opened their seventh production line, a 500 MW facility in Malaysia, in an effort to cut costs. But by waiting until 2009, the company was not prepared to immediately compete with the Chinese. In 2012 Q-Cells was purchased by Korean producer Hanwha Solar. The shift to Asia was a fitting illustration of the leadership transition in the industry. As one German interviewee put it,

> China had the right vision. When the world talked about 100MW, Chinese talked about 1GW. They had it right.

Q-Cells' fall is surprising given their position in the PV supply chain. They had a proven record of nearly a decade of world-leading capacity expansion. They had distinguished themselves from the aversion to risk that typifies many German firms. They had close access to the key material and equipment suppliers, as the biggest of the former, and most of the latter, were based in Germany. They had access to world-class engineering talent. They had some favorable production cost assets in that they could locate in the former East, where costs were lower and investment was still being subsidized. Crucially, they had a front row seat to policymaking and close contact with the most important PV market in the world. Somehow that wasn't enough. The two decisions above—expensive silicon contracts and a diversified technology strategy—contributed to Q-Cells losing its comparative advantage to Chinese firms (Figure 6.2).

Several other factors hastened Q-Cells demise. First, after 2008 and the global financial crisis, Q-Cells could not access capital. That was a major change. At that point, Q-Cells only had access to debt financing and its profit growth was too slow to convince bankers it could meet the payments. Once revenue fell in 2009, Q-Cells was shut off. In contrast, 2009 was the year in which the Chinese central government initiated an aggressive credit program for the PV industry (Chapter 7). The policy tool kit expanded, from support of deployment towards support of

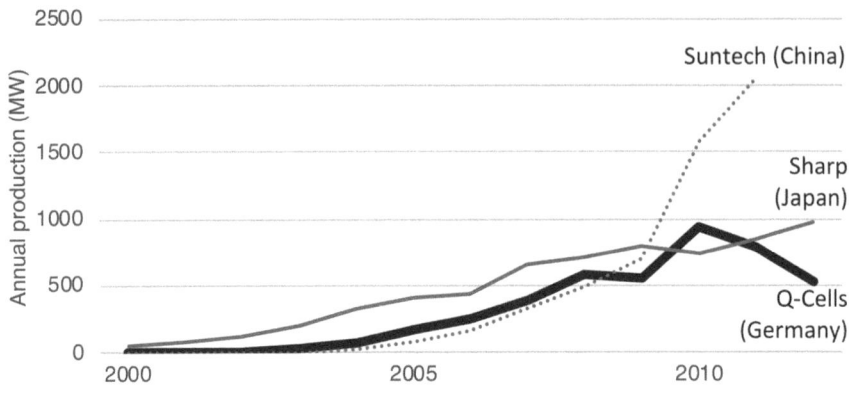

FIGURE 6.2 Annual production (MW) by Q-Cells, Sharp, and Suntech.

financing for scaling up production plants. This policy innovation was, however, not adopted in Germany (Grau et al., 2012).

Second, Q-Cells ran into constraints on human capital; it couldn't find enough scientists and engineers. Evidence comes in the offers to graduate students to drop out of school to work for Q-Cells. Those that were available could demand high wages contributing to Q-Cells high cost structure.

Third, Q-Cells, as well as Solarworld, paid very high prices for the latest manufacturing equipment. Since they were the first customers, these machines were custom made "on spec" and were expensive. German producers often received beta-type machines. This technological immaturity caused production delays and forced company resources, in the form of engineers' time, to be spent bringing these systems into production rather than scaling up. In contrast, those machines could be refined and standardized by the time they were sold to manufacturers in China making them cheaper and more mature. The Chinese thus had access to proven low-cost machines that Q-Cells did not, even though they had funded the development of them.

Fourth, interviewees pointed to systemic aspects of German corporate culture that apply even to a path-breaking start up like Q-Cells. One is the slowness of innovation in part due to lengthy testing, that is required when taking quality promises seriously. An example is that of a new silver paste product from DuPont for making electrical contacts on a PV cell. Bosch took 6–12 months to qualify it before it could be used in products. There is a caution that may be indispensable in the automotive sector but is not valuable in PV. In contrast, Chinese firms, while initially skeptical of a new technology, will rapidly adopt new production processes once they see some demonstration of its effectiveness in another company. The innovation will spread by word of mouth and information exchange leading to very rapid dissemination of innovative techniques. German adoption was further slowed by an inclination that each firm should invent things for itself rather than replicate success that can be observed elsewhere.

Making the machines that make PV

German equipment makers were among the first to supply machines to the PV industry. As markets grew, Centrotherm, Roth & Rau, Manz, Rena, Asys, Schmid, Innolas, Jonas & Redmann, and others were able to increase the throughput of their machines rapidly, which strongly contributed to the rapid growth of PV. Outside of Germany, Baccini (Italy), Tempress/Amtech (Netherlands/USA), MeyerBurger (Switzerland), GTAT (USA), Applied Materials (USA), and others produced equipment as well. These companies only became seriously interested in serving the PV market in the 1990s.

The German PV industry in the 1990s was a quiet place, never exceeding 10% market share in either production or installations. PV producers, even those owned by large companies, were small operations manually processing PV cells. Even in 1995–97, there was no specific PV manufacturing equipment. Most companies were using second hand equipment purchased from the semi-conductor industry. In 1998–99, PV manufacturing volumes were still quite low—a few MW for the entire country—not enough for PV producers to justify investments in specialty equipment. Due to the small size of the market, equipment providers could not justify building their own PV-specific machines. Instead, producers did what they could with second-hand machines from the semi-conductor industry and, in some cases, began to modify them to better suit the PV production process. Even outside Germany, in the late 1990s leading producers such as Solarex in the US and Kyocera in Japan were operating production lines that involved a series of machines with manual handling of materials and product between each one. Further, each machine was second-hand from companies that had used them for electronics.

By 2000, some in the industry had begun to observe this and a notion emerged that if PV was to become a real industry, it would need its own manufacturing equipment. The first machines specific to solar were probably from the US with Tyco's edge-fed growth process to manufacture ribbon silicon. In the 1980s, HCT Cutting Technologies of Switzerland developed a wire saw that was a critical breakthrough. It used the wires developed for radial tires and could saw 4,000 wafers simultaneously. This was an exception, a rare example of an innovation originating in PV and that was then utilized within the chip industry.

In other areas—such as texturing cells to minimize reflectance and stringing machines to electrically connect cells—PV specific equipment was needed because, like ribbon silicon, an equivalent process did not exist in semi-conductor manufacturing. Even so, some of these, such as stringing, remained manual processes into the 2000s.

One early effort was GP Solar, based in Konstanz, which began to produce testing equipment for quality control of cell lines. This was a new type of equipment and was, crucially, specific for PV production, rather than repurposed equipment from the computer industry. Around the same time, Centrotherm, which had traditionally made furnaces for the semi-conductor industry, began to focus its product portfolio on PV. Diffusion furnaces are used to diffuse dopants into the surface of silicon wafers and form a p–n junction—for example by having the surface absorb phosphorus from gaseous phosphoroxycloride ($POCl_3$). Centrotherm had been making diffusion furnaces for the semi-conductor industry for years and it began to develop a diffusion furnace specific to PV. Because the homogeneity requirements for PV were much less stringent than for semi-conductors, Centrotherm realized that many more silicon wafers could be stacked inside a furnace. The wafer surfaces would still be coated with phosphorous silicate glass, just not as uniformly. In this case, they needed only a simple modification: new wafer holders (called boats) that could fit many more wafers in one furnace process, and eventually longer tubes with bigger diameters. From there it was a process of continual adjustments and upgrades to produce a reliable PV-specific diffusion furnace with high throughput. A breakthrough in this process was by Jonas & Redmann, who automated loading of the quartz boats with 200, 400, and later 1,000 wafers.

With the EEG revisions in 2004, the German market went from 25% of global installations to over 60%. This explosive growth generated both the need and opportunity for PV-specific equipment. For example, until 2004, the PV industry used two key sources of off-spec (lower quality than could be used) material from the computer industry: purified silicon and wafers. With the EEG, the supply of off-spec materials was insufficient. Thus began the silicon supply crisis and the need for alternative suppliers.

The silicon manufacturing industry consisted of an oligopoly of five large producers who enjoyed the rising prices and were reluctant to expand production. In addition, it had faced overcapacity issues in the semi-conductor industry just a few years before and were cautious to avoid another overcapacity cycle. Silicon processing required technical knowledge and capital-intensive equipment. These firms were making silicon for $25 per kg and selling it for over ten times that price. One of the largest, Wacker, had considered working on a PV-specific program in the 1990s but the family that controlled the business deemed the market too risky (Palz, 2010). Firms began to sniff around and enter the silicon industry. One startup hired engineers from MEMC and Hemlock and decided to produce polysilicon specifically for solar. It would be less capital intensive than computer industry grade and would be a factor of 100 less pure, which would cut costs by a factor of two. Given how high prices were it was not hard to attract investors; raising $100 million was feasible. Within four years, this approach to making silicon had become widespread and prices crashed.

The rapidly growing market also began to create expectations that there could be substantial returns to investments in developing PV-specific machines. German PV companies such as Q-Cells, ASE, Solarworld, Nukem/Schott began to invest to meet the rapidly expanding demand. The first PV manufacturing plant with PV-specific equipment was Solarworld's production line in Freiburg, which came on-line in 2004, just in time for the EEG. It was the first plant that would look generally similar to a PV plant in 2018. Sharp, the biggest producer in the world at the time also built a modern factory that year, as did Kyocera. Each installed PV specific equipment for plasma enhanced chemical vapor deposition (PECVD), diffusion furnaces, screen printing, gas coating, and quality control.

PV producers paid for the development of these machines and helped to debug them. Early funding for the equipment manufacturers came from national R&D sources in Germany, Switzerland, France, the US, and Japan. But equipment manufacturers ultimately used cash flow from sales to PV manufacturers to finance the development of these first PV-specific machines between 1998 and 2004. All machine makers adopted a similar process. First, they would design a customized piece of equipment for a customer. They would charge a high price for that first model that would cover, at least partially, the cost of the expensive first-of-a-kind machine. Essentially, the first customers were paying for the development cost. And these customers were taking a risk that the first machine would be reliable and effective, and contributed much to the development cost by devoting their in-house engineering staff to debug the systems. Even when interest from Chinese customers in these products increased, European equipment manufactures would favor a European pilot customer, because communication was easier, travel shorter, and support by engineers better.

Despite the costs of early adoption, there was some assurance that came from the machine manufacturers' experience in semi-conductors—not only with the technology, but in terms of managing customer expectations and being available to service and adjust the first versions of a machine put into operation. The turnkey approach that emerged later made performance guarantees an essential strategy to encourage newcomers without experience to enter the market. As the EEG took off in 2004, having access to state-of-the-art production technology that enabled higher throughput and high yields became crucial, so PV producers were willing to take the risk on new machines. Once financing became abundant after 2004, they could afford to pay for the development cost, at least in the near term. Often the development cost was not too high since these firms were not starting from scratch but rather, modifying a piece of semi-conductor equipment.

A wide array of machines emerged from this financing model. Meyer-Burger developed wire saws to cut many wafers from an ingot simultaneously. HCT used machines for sapphire cutting, which was used as a substrate material in the display industry. Wacker developed a Siemens reactor for chemical vapor deposition of poly-silicon. PVA Tepla developed a mono-crystalline wafer puller that was specific top PV but which originated in its long experience in producing high purity silicon for the computer industry. Crystal Systems developed poly-crystal machines in the US. Some of these firms were quite focused on keeping competitors out of their markets for their machines.

Their customers, the PV producers, financed these machines with a variety of methods. PV firms owned by oil companies received allocations for capital investment from their mother companies. Solarworld was privately financed and used some of that funding for equipment purchases. By 2005, initial public offerings emerged as a major source of capital, for Q-Cells in Germany, as well as Suntech and others in China. In Figure 6.3 one can see the rise in R&D by machine makers once the EEG was in place and well established (Breyer, 2012). By 2009, equipment makers accounted for 7 of the top 15 solar companies ranked by R&D intensity (R&D/sales).

The idea of turnkey systems emerged in 2001 and 2002. The US semi-conductor equipment manufacturer, Applied Materials was in this space early, but with thin film silicon technology, which disappeared after a short climax quickly. Centrotherm delivered the first turnkey PV line to Solarworld. Solarworld founder, Frank Asbeck knew he wanted to build a PV cell factory but did not have the technical capability. Centrotherm hired engineers who had been involved in building a line at Sunways. In 2004, Centrotherm made a proposal to Solarworld for the "Deutsche Cell," a turnkey line to produce cells from wafers. Centrotherm would supply the equipment and set up the line. Their proposal included performance guarantees for cell efficiency, throughput, yield, and cost. Centrotherm intended to be judged in these specific metrics. The deal involved an up-front lump sum payment.

It soon became clear to other PV manufacturers that Centrotherm could set up an entire line well. They soon received an order from the other major German PV startup, Q-Cells. In this case, the arrangement was for a lighter version of "turnkey," in which the customers procured some of the equipment themselves, while Centrotherm supplied its own components and set up the line. Centrotherm began selling turnkey lines to customers in Taiwan, then Korea, and later China. It soon had 80–90% of the market for turnkey cell lines. This opened the door for other equipment suppliers such as Applied Materials and Schmid to provide turnkey solution equipment for Taiwanese, Korean, and Chinese fast followers to rapidly enter the global solar PV industry.

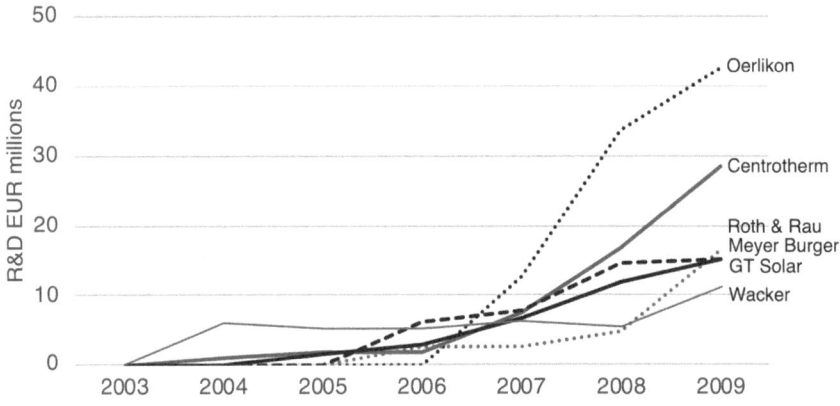

FIGURE 6.3 R&D investment by PV equipment makers (millions of euros).

One supplier perspective is that to be successful in turnkey, the customer needs to be able to jump to the top of the industry on technology. It allows them to become higher-tech, but not so much that it is risky. Suppliers sold turnkey on the premise that it takes away the risk, due to install expertise and performance guarantees. If the targets are met and the customer was happy, the expectation was that they would purchase more machines from that supplier.

A common technology platform emerged that allowed equipment suppliers to make use of their expertise outside PV, such as in semi-conductors, and thus made the interlinking of components more feasible than it had been before (Wu and Mathews, 2012). Only rarely do turnkey lines provide a start to finish solution. Rather, turnkey typically refers to automating one step in the value chain (polysilicon, crystal, wafer, cell, module). For example, turnkey solutions could produce cells from wafers (cell factory) or produce modules from cells (module factory). One study from this period showed that 74% out of 111 equipment suppliers "only provide equipment for one production step" and only one of the companies provides equipment for all steps along the value chain (Neuhoff et al., 2007).

Nevertheless, with the advent of turnkey equipment options, more firms could enter the market without the know-how that was imperative to the success of the early PV firms. This became especially important for some firms in China after 2005 when markets grew rapidly and financing was plentiful. Companies that opt for turnkey solutions could focus on production rather than R&D. Turnkey allowed firms to substitute money for know-how. Over $30 billion flowed from Chinese suppliers to European equipment suppliers. There is some evidence that those who opted for turnkey equipment often struggled with profits, for example due to low-module efficiency (Schmidtke, 2010). Eventually, Chinese customers preferred buying single equipment from the equipment suppliers and sourcing technical know-how for process integration and optimization elsewhere, e.g. from German manufacturing process consultancies. Turnkey equipment was more expensive and costs were such a prominent concern. In 2018, only a minority of firms offered turnkey solutions (Table 6.1). Still, turnkey offered speed, which was highly valued in China. Once equipment was procured and a line was set up, PV producers could focus on what they do best—optimizing yield and gradually increasing the electrical efficiency of their product.

PV manufacturing equipment started going to China in a big way from 2010 onwards when the market was growing fast and financing equipment purchases was relatively easy. In 2000–05, most machines were delivered to firms in Europe but by 2010 over 90% went to China. These machines included processing equipment, surface cleaning, and various stages of the manufacturing process of cells. There was less on cell-to-module processing, which was manual for a surprisingly long time until it was eventually automated by Chinese machines.

One interviewee estimated that $30 billion of machines went from Europe and the US to China between 2008 and 2012. Demand did not steadily grow since the machines were for lines that are typically lumpy investments. There were some down years. Many of the companies who sold the equipment to China would send

TABLE 6.1 Global manufacturing equipment providers in the PV industry in 2018.

	Ingot	Wafer	Cell	Crystalline Panel	Thin Film Panel
All firms	88	318	479	617	219
Firms offering turnkey equipment	5	9	17	56	17

Source: ENF Solar (2018).

technicians to Chinese factories to set-up, service, and provide training on the equipment they sold. Germany was setting China up to be successful PV manufacturers, hoping to gain from the knowledge and equipment transfer they were providing.

Eventually the Chinese begin to make their own machines, and they did so quickly. They didn't have the same reliability at first. In 2004 many Chinese firms didn't need to know anything about solar cells, they just needed the money to buy the equipment that does it all. As one interviewee put it

> I was part of a German consultancy firm that helped Chinese companies set up production sites sometimes even with the knockoffs. We would go to China to set up equipment and teach people how to use it. I went on a trip to Solarfun and Trina, backed by a Swiss venture fund that set up Solarfun. We were instrumental that in transferring tech know-how in Germany to China. Doesn't take much time to get it running and get it started to run at optimal performance to ramp up production. And once companies have the equipment, the production is easy to make. We could transfer know-how to 4 people within 2–4 weeks.

Reverse engineering was happening and, as a result, innovations in equipment spread quickly.

Some expect that in the next phase a few big machine makers will produce them in China and in the rest of Asia. New players are emerging who benefit from western machine making and address skepticism by demonstrating established German design and quality. However, they also benefit from producing in China, where Chinese design engineers are fast and pragmatic, and costs are still much lower. In 2018 turnkey was happening in India, Turkey, and the Arabian Peninsula. In Turkey what is known as turnkey light is used, in which producers there want to negotiate on their own for other equipment.

German electric ratepayers invested over $200 billion to create a market for PV that was far bigger than almost anyone had imagined it could be. Not only that, but it was big enough to catalyze a new industry—PV-specific equipment suppliers. The market was also big enough that investors and entrepreneurs began to target that market with consequential activities. The most consequential of these entrepreneurs was in China, where start-up firms began to meet the demand that German producers could not—or would not—meet on their own, eventually making use of that equipment to scale up to even larger levels of production and very large cost reductions.

References

Breyer, C. 2012. *Economics of Hybrid Photovoltaic Power Plants*, book-on-demand.de.
Creutzig, F., Agoston, P., Goldschmidt, J. C., Luderer, G., Nemet, G. & Pietzcker, R. C. 2017. The underestimated potential of solar energy to mitigate climate change. *Nature Energy*, 2, nenergy2017140.
Davidson, O. G. 2012. *Clean Break: The Story of Germany's Energy Transformation and What Americans Can Learn from It.* ebook.
DPG 1987. Gemeinsamer Aufruf der DPG und der DMG. Warnung vor weltweiten drohenden Klimaänderungen durch den Menschen. *Physikalische Blätter*, 43, 347–349.
ENF Solar. 2018. Available: www.enfsolar.com/. [Accessed 15 March 2018].
Gillis, J. 2014. Sun and wind transforming global landscape. *New York Times*, September 13, p. 1.
Grau, T., Huo, M. & Neuhoff, K. 2012. Survey of photovoltaic industry and policy in Germany and China. *Energy Policy*, 51, 20–37.
Hager, C. & Stefes, C. H. 2016. *Germany's Energy Transition: A Comparative Perspective*, Springer.
IEA 2006. *Energy Technology Perspectives – Scenarios and Strategies to 2050*, Paris, International Energy Agency.
IPCC 2011. *The IPCC Special Report on Renewable Energy Sources and Climate Change Mitigation*. Intergovernmental Panel on Climate Change (IPCC).
Jacobsson, S. & Lauber, V. 2006. The politics and policy of energy system transformation—explaining the German diffusion of renewable energy technology. *Energy Policy*, 34, 256–276.
Jacobsson, S., Sanden, B. A. & Bångens, L. 2004. Transforming the energy system—the evolution of the German technological system for solar cells. *Technology Analysis & Strategic Management*, 16, 3–30.
Krause, F. 1980. *Energie-Wende: Wachstum und Wohlstand ohne Erdöl und Uran*. Frankfurt am Main, Öko Institut.
Lovins, A. B. 1976. Energy Strategy: The Road Not Taken? *Foreign Affairs*, 55, 65–96.
Meadows, D. H., Meadows, D. L., Randers, J. & Behrens, W. W.III 1972. *The Limits to Growth*, Washington, DC, Potomac Associates.
Morris, C. & Jungjohann, A. 2016. *Energy Democracy: Germany's Energiewende to Renewables*, Springer.
Neuhoff, K., Nemet, G., Sato, M. & Schumacher, K. 2007. *The Role of the Supply Chain in Innovation: The Example of Photovoltaic Cells*. Cambridge, University of Cambridge – Electricity Policy Research Group.
Palz, W. 2010. *Power for the World: The Emergence of Electricity from the Sun*, Pan Stanford Publishing.
Quitzow, R. 2015. Dynamics of a policy-driven market: The co-evolution of technological innovation systems for solar photovoltaics in China and Germany. *Environmental Innovation and Societal Transitions*, 17, 126–148.
RESA 2001. Act on Granting Priority to Renewable Energy Sources (Renewable Energy Sources Act, Germany, 2000). *Solar Energy*, 70, 489–504.
Scheer, H. 2004. *The Solar Economy: Renewable Energy for a Sustainable Global Future*, Abingdon and New York, Earthscan.

Schmidtke, J. 2010. Commercial status of thin-film photovoltaic devices and materials. *Optics Express*, 18, A477–A486.

Staiss, F. & Räuber, A. 2003. Strategies in photovoltaic research and development—market introduction programs. In: Bubenzer, A. & Luther, J. (eds.), *Photovoltaics Guidebook for Decision-Makers*, Springer.

Steinbacher, K. & Pahle, M. 2016. Leadership and the Energiewende: German leadership by diffusion. *Global Environmental Politics*, 16, 70–89.

UNFCCC 2015. *The Paris Agreement. United Framework Convention on Climate Change.*

Unnerstal, T. 2017. *The German Energy Transition: Design, Implementation, Costs, and Lessons*, Springer.

Wene, C.-O. 2000. *Experience Curves for Technology Policy*, Paris, International Energy Agency.

Wu, C.-Y. & Mathews, J. A. 2012. Knowledge flows in the solar photovoltaic industry: Insights from patenting by Taiwan, Korea, and China. *Research Policy*, 41, 524–540.

PART 3
Making it cheap

7
CHINESE ENTREPRENEURS

In January 1994, Martin Green, David Hogg, and Zhengrong Shi flew from Sydney, Australia to China. Professor Green and his lab group at the University of New South Wales (UNSW) had recently developed their high efficiency solar cell and Green felt that it was time to commercialize it. Hogg was in charge of UNSW intellectual property and Shi was a post-doc in Green's lab whose role was to serve as translator in China. The trip was a disappointment. They were looking for a partner to set up production, but all the locations they visited were, in Green's words, "hopeless—devoid of all appropriate infrastructure." They returned to Sydney and started the company there instead. They got the local utility, Pacific Power, to invest $45 million and named the company Pacific Solar. The trip by Green, Hogg, and Shi was perhaps only a matter of months too early. The business environment in China would improve dramatically over the next four years. The same year of that first China visit, the central government implemented its far-reaching "tax sharing" reform. It raised tax receipts overall but, more importantly, it changed the allocation of tax revenues so that municipal governments could retain a much larger share of the tax revenues paid by companies within their cities (Lou, 2008). Suddenly, they had strong incentives to recruit businesses to locate in their cities. By the time members of Green's team returned in 2000, the business environment had been transformed.

Overview of China's contribution

Albert Einstein, Bell Labs, the US Block Buy, Japanese niche markets, and the German feed-in tariff (FiT) all played key roles in developing PV. But activities in China between 2000 and 2016 contributed most directly to cheap PV. During that period, Chinese solar companies scaled up production by a factor of 500. By 2007, China produced more PV than any other country and by 2013 it was installing more PV than any other country. By 2017, China produced 70% of the world's PV.

To understand the distinct path that China pursued in this period, I focus on one pioneering company, Suntech, founded largely by former members of Green's team. Almost all of those founders were Australian citizens, some by birth and others who in becoming naturalized Australian citizens had had to renounce their Chinese citizenship. Suntech was the seed from which the Chinese PV industry grew, flourished, and eventually dominated. It established a successful model that others followed. It catalyzed cities' interest in PV. It created a domestic Chinese supply chain. It trained a local skilled labor force. It partnered with foreign firms to export to foreign markets. It accessed finance from US capital markets. It established a profitable business model. Suntech demonstrates the role of entrepreneurship, which is a driving force behind innovation in China today. The Chinese central government's motivations for PV have covered a range of goals over time: national security in the 1970s, poverty alleviation in the 1980s and 1990s, economic growth post-2008, and environmental concerns in the 2010s. And the state certainly played a role in enabling a thriving PV industry. However, the state's role in PV was secondary to that of entrepreneurship in the 2000s (Quitzow, 2015). It wasn't until after 2009 that the state played a critical role in enabling the thriving PV industry.

This dichotomy is most clear in the shifting sources of funding for PV companies. During the 2000–07 period the Chinese industry went from a few nascent start-ups to the world leader in PV. Financing in the early period was scrappy and dispersed—a few million dollars here and there coming from individual PV entrepreneurs themselves and angel investors. Most companies started as Australian–Chinese joint ventures, with expertise coming from Australia, some start-up capital from China, and subsequently much larger investments from the US. The central government created a small domestic market with the Brightness Program in the 1990s and had supported some state firms to produce PV to power satellites and communications before that. But the Chinese central government only supported PV materially beginning in 2009. This was after firms had established the legitimacy of the industry with their Initial Public Offerings (IPOs) in New York in 2005–07. As in many aspects of PV in China, Suntech led the way. Chinese PV companies raised $7 billion from US capital markets from Suntech's December 2005 IPO until the end of 2007. The central government became strongly supportive in 2009 and it extended tens of billions of dollars in credit to PV firms in 2010. Though late, the government's participation was still important because capital requirements for scale up involved billion-dollar decisions. Global demand for PV then accelerated with annual growth in shipments increasing from 30% per year to above 50% per year.

The explosion in demand for PV set off by the German FiT in mid-2004 transformed Chinese production. The satellite programs of the 1970s and 1980s, and rural electrification initiatives in 1996 and 2002 had established a small Chinese market. But the German market was two orders of magnitude larger. Moreover, the planned reduction in rebates in the German policy meant that there was a rush to install solar before the subsidy program ended. During 2004, participants in the solar industry suddenly realized that they needed to move quickly. The Chinese are better at building quickly than anyone. The US, Japan, and Germany were simply not fast enough to recognize the market opportunity, much less meet it. One of the distinct characteristics of firms in China is their "organizational capability"—their competence in hiring quickly, producing quickly, and improving quickly. These qualities were exactly what the German PV market required.

China became serious about its own domestic market in 2005 with the passage of its Renewable Energy Law. As in other countries, wind power served as a helpful precursor technology for PV. While the law mainly targeted wind power, it bolstered the confidence of investors in Chinese PV companies by conveying an expectation that China would one day become a major market for PV. The Chinese FiT, created in 2011, allowed the industry to continue to grow even after the global financial crisis and reform of the German subsidies slowed demand for Chinese exports.

Firms in the Chinese industry contributed to the development of PV technology in a variety of ways. The combination of PV R&D, early PV production, and the emergence of the semi-conductor industry during the 1970s–1990s provided a foundation of competence. Experience in other areas of high-volume manufacturing, such as textiles, also played a role. From 2000, substantial know-how was imported, such as extensive international collaborations and repatriated Chinese who had been trained in the West, particularly in Australia (Binz et al., 2017). Australia provided a steady stream of talent to Chinese PV firms, including individuals born in Australia but particularly Chinese-born individuals who spent many years there and in some cases had become Australian nationals. In addition, foreign machines—and the technicians who traveled to China to install and optimize them—provided steady inflows of knowledge to China. People that moved from Australia, Germany, and US to China were also crucial in China's success. Many firms upgraded gradually, starting out by producing modules from cells, which was the least technically challenging step.

Chinese firms achieved a key milestone when they obtained technology standards and certification, which provided reliability to skeptical early adopters in Germany. As Germany's market grew beyond the capacity of German firms, German distributors turned to China for modules. The technology standards helped convince the German's to take a chance on recently established Chinese firms. The German firms' rapid adoption of Chinese modules legitimized the Chinese technology.

Low labor costs gave China an international comparative advantage in the key scale up period of 2000–07. However, this advantage was short-lived due to the swift proliferation of automation. The labor share became too small to provide much advantage. To remain competitive, China relied on other advantages such as speed, flexibility, and rapid exchange of information among supply chain partners. At the same time that automation was increasing, costs were plummeting. As a result, experience outside of the semi-conductor industry became increasingly relevant. Producing textiles, shoes, and handbags provided know-how about producing at gigantic scale in a commodity market where margins were thin due to fierce competition. China had more experience with those activities than any other PV producer.

As the Chinese industry grew, it became extremely competitive. The global market share of the largest company declined and never exceeded 10% (Figure 7.1). Contrast that to Sharp's 28% of the world market in 2004. As competition intensified, profits were hard to come by, and costs continued to fall. China timed its scale up just about perfectly; in time to meet demand but not too early. There were times of over-capacity but that capacity was soon put to use in a global market that was growing at well above 30% per year.

The beginnings of PV in China

China has performed research on PV since the late 1950s. Since the Vanguard-1 launch in 1957, PV attracted intense interest among countries with space programs. China plunged into the space race, seeking to follow the Russians in launching a satellite into Earth's orbit. China launched its first satellite with PV, Dongfanghong-2, in 1971. China then became interested in R&D for terrestrial PV in the early 1980s (Zhi et al., 2014), launching massive rural electrification programs. PV's share of these programs was modest: the 6th Five-year Plan (1981–1985) included $10 million for PV R&D on efficiency improvements and cost reductions. Much of that was used to import seven turnkey lines from the US (Zhang and White, 2016). To put this scale in perspective, the US Budget during that period was 100 times larger.

Early production

In the mid-1960s, China established a small computer industry of state-owned enterprises that made semi-conductors. Two of these firms got into the solar business in the 1970s to fill demand from military communications applications, as well

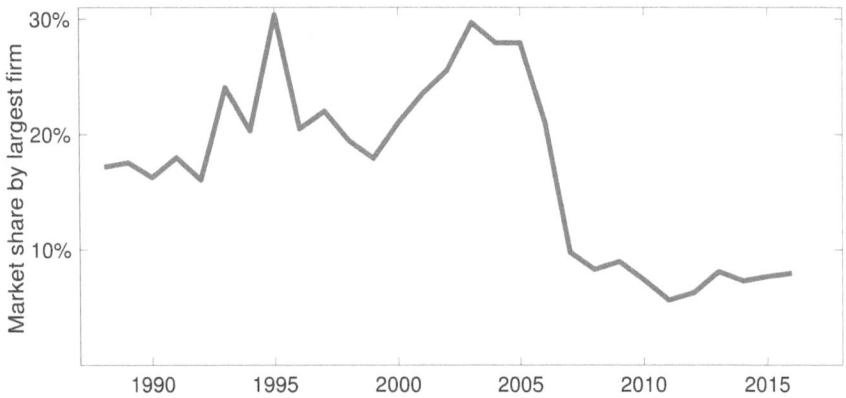

FIGURE 7.1 Share of global production by the largest firm in the world.

as satellites. Kaifeng Solar Cell Factory in Henan started PV research in 1970 and began making mono-crystalline PV cells in 1975. The Ningbo Solar Power Source Factory began making cells and modules in 1978, and the Yunnan Semiconductor Devices Factory began making mono-crystalline cells in 1979 (Dunford et al., 2013). In the early 1980s, Kaifeng and Yunnan scaled up by importing production equipment from the US (Zhang and White, 2016).

Yunnan made particularly strong progress in that it was able to export its modules to European markets in the 1990s. This distribution channel was not a simple feat, and the volume was tiny. The Europeans were skeptical of the quality and reliability of these nascent Chinese producers and had imposed high hurdles for getting products certified in Europe. The sum of Chinese production was 0.5 MW in 1990 growing to 2 MW by the end of the decade. To put this in perspective, this meant growing from about 100 residential rooftop systems in 1990 to about 400 homes by 1999, in a county with 350 million households. Still, the semiconductor firms who established R&D developed an understanding of working with silicon and of producing at scale, which sowed important seeds for the PV supply chain that followed. They also established a set of personal relationships that would later become important for making things happen quickly at increasingly larger scales.

Early market

Rural electrification continued as a priority in the 1990s and solar played a role, albeit a small one. The Brightness Program, launched in 1996, was an initiative to use renewable energy to provide power to people in rural western China. The target was to use renewables, particularly small-scale wind power and rooftop PV, to provide electricity access to 23 million people by 2010. It turned out, however, to be an expensive way to provide power. The 2003 follow-up program, the SongDian Dao Xiang (Township Electrification Program), was larger with funding of more than RMB 3 billion and installation of 20 MW of PV. PV did not play a major role in electrifying households in either program. However, the programs were important to PV's growth in China. Installed PV grew from about 5 MW in the mid-1990s to ten times that by 2004. Companies from the US (BP and Shell Solar), Germany (Siemens Solar), and Japan (Sharp and Sanyo) began to establish a presence in China. A new set of domestic firms started to play a role as well. Throughout this period the Chinese solar market remained minuscule due to the high costs of PV and the widely held perception that it would stay that way.

International cooperation

In the early 2000s, China began to engage in an array of international collaborations on renewable technologies. There were bilateral cooperation agreements on wind power with Denmark and Germany. The World Bank and the Global Environmental Facility became interested in funding renewables projects in China.

PV-specific collaborations included the China–Japan New Energy and Development Technology Organization Project, the China–Germany Western Solar Energy Project, as well as bilateral cooperative agreements with Canada and the Netherlands. Some of this exchange was technology based. But some of it was policy exchange. For instance, in 2004, as the 2006 Renewable Energy Law was being planned, government officials traveled to Denmark, the US, Spain, and, most importantly, to Germany to learn about implementation issues in their support policies for renewables. By 2010, China had signed over 60 international cooperative agreements (Zhao et al., 2011). These formal arrangements facilitated the direct exchange of informal knowledge exchange—the know-how that moved globally as scientific publications, as tacit knowledge in people, and as knowledge embodied in machines.

Government support

Chinese public institutions supported the development of the PV industry at multiple levels with a variety of instruments. However, the central government did not get directly interested in PV until late in the game. Wind had been the focus of renewable energy activity because PV was considered too expensive. That changed in 2005 and 2006 when Chinese PV companies began to list on western stock exchanges and bring in hundreds of millions of dollars of international capital to China. Coincidentally, the development of the 2006 Renewable Energy Law in 2004 and 2005 set expectations that China would soon—or would at some point—become a major market for renewable energy. After Chinese PV firms demonstrated legitimacy with their IPOs, "new energy" became a strategic industry in the 12th Ten-Year Plan. PV was a central part of new energy. That designation conveyed expectations to provinces and municipalities that there would be support for the industry from the central government.

In contrast, provinces and municipalities have actively supported the development of the PV industry, just as they have supported a wide array of other manufacturing enterprises. A 1994 tax reform allowed municipalities to retain a much larger portion of corporate tax revenues from companies within their jurisdictions. This likely set off a competition to attract firms and invest in their success. Municipalities offered free land, easy permitting in industrial parks, low-priced electricity contracts, helped recruit and train labor, provided temporary local tax credits, and a small amount of direct local financial investment. Municipalities occasionally co-signed loans, reducing the repayment risk for banks, allowing firms to borrow at lower interest rates.

Some of this local support was crucial in the early pre-IPO period. The IPOs brought legitimacy and large amounts of capital. Later, in the 2008–11 period, the investments required for scale up became much larger and state-owned banks became increasingly involved. For example, in 2010 China Development Bank provided a $9 billion line of credit to LDK Solar, a $5 billion loan to Yingli, and a $4 billion loan to JA Solar (Grau et al., 2012). Once the German market stopped growing in 2012, Chinese banks' willingness to restructure and forgive loans appear to have played a role in enabling some of these firms to survive.

Beyond this variety of forms of government support, culture also seems to have been important. Multiple interviewees in China pointed to "organizational capacity" as being a key enabler of the rapid scale up. It had to do with experience in other industries that made Chinese confident in their ability to set up infrastructure for new production. It involved coordinating land, money, and people. Most important, it was the ability to establish new facilities for production with speed. Another aspect of China's national innovation system is its openness to knowledge flows from abroad, a distinct contrast to the national innovation system of Japan, which was much more closed. Expertise from individuals trained in Australia was central to the establishment of the early PV industry in China. One firm brought both of these attributes—organizational capacity and openness to knowledge—together. The seeds of the China PV industry we have today were almost completely attributable to the founding and growth of Suntech.

Suntech

In Fall 1986, Shi Zhengrong had completed his master's degree from the Shanghai Institute of Optics and Fine Mechanics and applied for a scholarship that would allow him to do his PhD in the US. He was awarded the scholarship but instead it sent him to Australia, a country he would later admit he could not point out on a map. This was the same Mr. Shi who would later go to China with Martin Green in 1994.

Shi left Shanghai in March 1988 to attend graduate school at the University of New South Wales. His scholarship helped get him to Australia but it was not enough to live on in Sydney. His advisor in the physics department at UNSW did not have the funding that Shi really needed. His advisor told him to go to the electrical engineering department, "There is a guy there who is rich he has so much funding; go see Professor Green." Shi knocked on Green's door. He asked if he could get a job with him. Green apologized and told him that there are no positions available. Shi would not take no for an answer and told Green he didn't need a full-time job, he just wanted a part-time research position. The next day Green agreed that Shi could work in his lab. Three years later, Shi completed his PhD in 1991 in solar power technology with Green. He continued to work in Green's lab spending much of this time doing research on thin film PV-silicon on glass, working on ways to reduce the costs of modules.

China to UNSW

Shi was not the first Chinese student in Green's lab. Soon after he became leader of China in 1978, Deng Xiaoping launched a program to allow one thousand Chinese students to study abroad. China paid for their travel, tuition, and modest living expenses. One of this first cohort of 1,000 international students and researchers was Jiqun Shi who arrived in 1980 to study at UNSW. He found a position in Martin Green's solar power technology lab as a visiting researcher and quickly made a positive impression. Jiqun had been trained in micro-electronics in China and spent most of his time processing semi-conductor materials. Green was struck by how his training in micro-electronics enabled him to get up to speed quickly. Importantly, Jiqun brought "external knowledge" to the lab, which Green valued highly. Thus, in the early 1980s, a conduit for knowledge flow between China and Australia had been established.

After finishing his bachelor's degree, Jiqun went back to China. Soon after he did, Green received an invitation to visit. Green visited the Huazhong University of Science and Technology in 1984 and gave a 3-week course on PV based on his recently published text book (Green, 1982). Green was skeptical that the hundred or so students who packed the classroom each day were really interested in PV. It seemed more likely they had attended out of strong encouragement from those at the university who wanted to show respect to their visitor. Perhaps the most important outcome of the visit is that it encouraged Xi Chen to meticulously translate the new textbook into Chinese.

Green continued to visit China, giving talks and attending meetings there throughout the 1980s. He began to see development of cheap PV as a race against the impending industrialization of developing countries. If PV could get cheap fast enough, big developing countries like China and India could industrialize with PV rather than with fossil fuels. This urgency focused the lab group's attention on PV applications in China and India. Green's group had licensed high efficiency cell technology to India in the late 1980s and had helped set up a production line for this technology in New Delhi, but left convinced that it would be increasingly more difficult than establishing a Chinese operation.

Green's second Chinese student, Jianhua Zhao arrived in 1984 to do a PhD. Zhao later brought over his wife, Aihua Wang who also did a PhD with Green. Wang began working in the Green's lab processing high efficiency PERC cells. Zhao and Wang returned to China in 2006, as Vice President of Technology and Chief Technical Officer (CTO), respectively, of China Sunergy, the second company after Suntech to establish a benchmark 30 MW of cell production capacity in China (Green, 2016). Green's third Chinese PhD student, Ximin Dai also returned to China as CTO at JA Solar in 2005, the fourth company to reach this benchmark. The fifth Chinese student to arrive at Green's lab was Zhengrong Shi.

Pacific Solar

In 1994, Martin Green raised that $45 million to found Pacific Solar and commercialize the silicon thin film technology he and his students had developed in his lab. Green assumed the role of Research Director with David Hogg appointed as Managing Director. Shi, by then a post-doc, became Deputy Research Director and Director of Process and Technology Development. Those three had made that fruitless trip to China six months earlier. They set up a small production facility in

Botany, a suburb south of Sydney. In addition, Ted Szpitalak, who had been with Green since the early 1980s, joined with the crucial task of acquiring production equipment. His strength had been in acquiring and commissioning second-hand equipment. The source for all that equipment was the semi-conductor industry in the US. Green had given Szpitalak free rein to visit the US as much as he wanted in order to purchase this useful second-hand equipment. Szpitalak did so throughout the 1980s, establishing very strong contacts within the PV sector. Pacific Solar was focused on developing thin film solar technology. Szpitalak thought that focus was a mistake since all of Green's international success had been based on setting world record efficiencies in crystalline silicon. Shi also thought the thin film focus a challenge. As Deputy Research Director, Shi proposed to the board that Pacific Solar use some of its funding to build a crystalline silicon line. The board rejected that plan, concerned that established crystalline silicon producers, such as BP Solar and Sharp, had an insurmountable advantage in scale. By the end of the 1990s Pacific Solar needed fresh investors. Q-Cells bought its assets in 2004. However, since Pacific Solar had recruited many researchers from China to meet its staffing needs, its impact as a developer of human capital for the Chinese solar industry that followed dwarfed its direct impact on technology development and production (Green, 2016). Employees of Pacific Solar learned how to acquire equipment cheaply, integrate a disparate set of machines into a well-functioning line, albeit at small scale, and improve production by continuously tweaking it. David Hogg had also ensured staff were trained in modern management practice, by gaining ISO certification of the company's procedures and operations.

Business plan

While at Pacific Solar in the late 1990s, Shi noticed that PV policy support was emerging in Japan and Germany. He also knew that some of the early PV patents issued to Spectrolab—a producer of high efficiency space applications—were about to expire, making that technology available for free. He became increasingly interested in creating a commercial product. In 1999 Shi was visited by Huaijin "Sammy" Yang, a Chinese-born Australian citizen, who was enthusiastic about the opportunities created by the economic reforms that were happening in China. Yang would later become CEO of JA Solar. Yang encouraged Shi to look into starting a company in China.

In 2000, Shi, Yang, and Szpitalak visited China for two weeks, speaking with many people about establishing a PV business. As in their 1994 visit, Shi got the sense that it was still too early to produce PV in China. Only two PV companies were active in China, one producing 2 MW per year and another doing 200 kW. In total, Chinese production accounted for about 500 rooftop systems annually. The entire country's production was barely above laboratory scale production. However, the three found encouragement and at one point were asked to produce a business plan.

After the trip, Shi returned to Australia to write that business plan. His first task was to re-learn Mandarin, which he had stopped writing since moving to Australia 14 years earlier. He spent two weeks writing a 200-page business plan in Chinese. He realized he did not know how to calculate the costs of the materials and equipment he would need. But he intimately knew how his own lab worked. So, instead, he used the costs of those items in his lab and applied a 30% discount to account for savings from large-scale production. He calculated that if he could sell modules for $3 per Watt he could generate a 25% gross margin. The average module price in 1999 was between $3.80 and $3.50 per Watt.

Shi then presented his business plan to as many interested people in China as he could find. He went to Shanghai, Dalian, and Hangzhou, but found less interest than he expected. The one audience that did seem interested was the city of Wuxi, where municipal government employee, Mr. Jiajun Wang, the Deputy Secretary of Wuxi CPC, was keen on the idea. Wuxi wanted to attract high technology industries and figured they needed a high-profile professional to lead them. Dr. Shi was internationally established, well connected, and had scientific credibility. The Wuxi government invited Shi to come give a presentation.

In August 2000, Shi spent four hours presenting his ideas, answering questions, and discussing the possibility of establishing a PV company in Wuxi to an excited audience. One city official told him, "I want you to be the general manager, not the CTO." Shi says that that was the first time he envisioned himself as a CEO rather than as a technology lead. However, a crippling obstacle remained. Despite their enthusiasm, it was quite challenging for Wuxi government to arrange the investment to found the company.

Meanwhile, Sammy Yang had been traveling throughout China to find funders. After nine months, Yang had gotten nowhere. No one was interested. In the end, it was Colonel Jiajun Wang who exercised his administrative influence on some local companies. He called five companies in Wuxi and asked them to contribute a million dollars each for ownership stakes in a new solar company. These firms had nothing to do with PV, they manufactured other products, mostly appliances. Each one refused. However, the colonel was in a position to refuse to take no for an answer. The army's budgets had been cut in the 1990s and it had developed its own manufacturing businesses to fund its operations. And since the army controlled the five companies the colonel had called, he made a stronger request. The next morning an account had $6 million in it.

Before closing the deal, the Wuxi government had a third party conduct due diligence on Shi's business plan, inspected his home in Sydney, and interviewed his colleagues at UNSW. The final consortium included two venture capital investors—Wuxi Venture Capital Group and Wuxi High-tech Venture Capital Company—as well as the investment arms of the five local corporations: Wuxi Keda International Investment, Wuxi Guolian Trust and Investment, Little Swan Group, Wuxi Mercury Group, and Wuxi Shanhe Group (Hejun, 2015). Shi contributed the $400,000 he had raised on his own, which gave him a 5% stake in the company. He also assigned to the new company 14 PV patents he held and received an additional 20% stake for those. The seven state owned enterprises owned the other 75%. With that $6 million behind him, Shi knew that he could quit Pacific Solar.

But Shi was not willing to drop Pacific Solar right away. He had earned the trust of Green over a decade and a half and was anxious about the idea of letting the group down. He gradually leaked the news of his coming departure over three months in the Fall of 2000. Most important, Green was supportive. Green told Shi, "I support you 100% on this. I know this is something you have thought through carefully so I am sure you are doing the right thing." Green says that the turning point for him was when Zhengrong told him that, if he did not soon take his two Australian-born sons back to China, they would forever be estranged from Chinese society. The deal for the establishment of the new company, Suntech, closed in January 2001.

Line #1: start up

In March 2001, Shi, his wife, and their two boys prepared to leave Sydney. They packed up their PV research literature into 20 boxes and sent them on a ship from Australia to China. Then all four of them flew together to their new home in Wuxi.

Once they arrived, Shi and Szpitalak realized they did not know much about doing business in China, particularly regarding the availability of equipment and supplies. For example, they initially sourced their glass from Australia. As described in Shi's business plan, the first step was to use the $6 million raised to establish a manufacturing line that could produce 3 MW of modules per year, a "3-megawatt line." The line began operation on September 9, 2002. Through their creative use of the resources available, they were able to produce at a rate of 10 MW of modules by the end of 2002, the first year of Suntech's production.

Shi felt that running a production line well and keeping on top of the rapid technological improvement in the industry meant you needed to do research of your own. Accordingly, and despite the extreme resource constraints, Shi used $1 million to build an R&D laboratory, which later proved to be key in keeping Suntech at the forefront of a rapidly improving industry. That was the first PV lab in China built to international scientific standards. They designed everything to use the most advanced technology, even as they took every possible opportunity to keep operations cheap. Szpitalak scrambled to buy second-hand equipment from other PV companies, as well as from the semi-conductor industry in China. It was also an early indication of Suntech's challenges in being a PV pioneer in China. It would have to establish a supply chain on its own—over and over they would do things for the first time in China.

Having moved his family to Wuxi, Shi hit the road. He paid for Szpitalak and himself to fly to equipment manufacturers in the US, Japan, and Germany in order to find equipment for their 3 MW line, which they were in the process of upgrading to 10 MW. Shi and Szpitalak built the 3–10 MW module production line in six months and took 12 months to build the cell production line. Suntech recruited production engineers who had experience working in the semi-conductor industry in China. After three months of working 16 hours a day they were able to start producing small amounts of modules with efficiencies of 15% for multi-crystalline modules, which was slightly above the industry average at the time. This high efficiency helped establish Suntech's reputation in China; they were making a world class product. In addition, Green convinced Shi that it was necessary to communicate to the rest of the world the high quality of these modules. Shi pursued International Standards Organization 9001 certification, which no other company in China had. Shi also spent the money to seek certifications from the International Electrotechnical Commission and the European Conformity quality standards, which they obtained in 2003. Shi knew solar markets well enough that from the outset they were producing goods for export, most likely in Japan (the biggest market at the time) and possibly Europe. But so far, Suntech was only selling modules in China. In fact, four senior managers left at this point in the company's development, concerned that the Chinese market was too small and that European markets would remain inaccessible (Zhang and White, 2016).

By the end of 2002, Suntech was just getting started. It was not quite hitting the $3 per Watt cost target that Shi had sketched out in his business plan two years earlier. But they were not far away. They were producing modules at $3.28 per Watt and were selling them at $3.48 per Watt. Still, while they were able to produce above variable costs, Suntech was not making a profit, losing about $1 million that year (Hejun, 2015). But Shi was focused on the long-term potential of these modules, and that they could be sold above the costs to produce them at this early date was encouraging.

Line #2: Germany

The departure of the senior staff in late 2002 was unsettling and caused Shi to shift his emphasis from technology and production to marketing. In March 2003, Shi attended an international solar conference in Berlin, Germany. Suntech was the only Asian company there—the world leader at the time, Sharp, did not attend. At first, the attendees looked at Suntech's booth dismissively. They seemed surprised when an Asian man spoke to them in almost perfect English. Mentioning his work as a student and researcher at UNSW gave him instant credibility with the Germans, as the university's steady stream of efficiency world records was well known. They appreciated the engineering prowess. The skeptical Germans soon gave the guys in the Suntech booth a chance to prove themselves. Suntech sold out its full annual output of their 10 MW year line in the next two months. Shi's pivot to marketing had worked—now they needed to expand.

But they had already burned through their $6 million initial investment by the local Wuxi manufacturers. Shi appealed to the Suntech board to expand production. As the Wuxi government had done two years prior, the board gave their approval but did not offer any financing. The local banks noticed that Suntech was selling all the modules they could produce, and that they were able to sell them above their production costs. Suntech began to get loans from the local Chinese banks, receiving $5.5 million from the Wuxi government between 2003–04.

After a couple of years in China, Shi discovered that, despite the absence of PV companies in China, there was some emerging related activity. Suntech had built their first 10 MW production line using silicon wafers made from poly-crystalline silicon in Germany. Mono-crystalline silicon tends to have slightly higher efficiencies because it does not have losses associated with the edges of crystalline regions that poly-silicon suffered from. But it is more expensive to pull a single crystal out of molten silicon than it is to form thousands of crystals in a mold. However, in China, manufacturers were already making single crystal ingots for the memory-chip and micro-processor industries. Shi found they could source cheaper mono-crystalline silicon in China than poly-crystalline from Germany. He made a somewhat bold cost-reducing move and decided that line #2 would use a different process from line #1 so that they could use mono-crystalline wafers made in China. Shi's experience in the lab and at Pacific Solar had given him a broad expertise in PV technologies that allowed him to switch technologies quickly when an opportunity for cost reductions emerged.

Still, getting the equipment to set up line #2 would be a challenge. Switching to the mono-crystalline process required passivating the cells with titanium dioxide liquid and that involved an expensive tool, one they needed quickly. Shi looked abroad to his international network of solar technicians to find it. In August 2003, he visited his friend Francesca Ferrazza at Eurosolare in Rome. She had one that had cost half a million euros, but sold it to him for 50,000 euros. Shi agreed on the spot and asked them to fly it to China immediately. But it was August in Italy, vacation time, and no one was in a hurry to pack up and ship an expensive and heavy piece of machinery to China. Shi returned to China with neither the machine nor his luggage as the airline had lost it. He called up his friend Francesca and asked where the tool was. She said the price had risen to 70,000 euros. He agreed and eventually received his machine soon after.

Shi faced similar challenges in acquiring a screen-printing machine, which cost $1.5 million a piece. Again, his international network paid off. He knew that a Japanese company, NPC, wanted to start making screen-printing machines. He offered to give them their first order. He would take their first machine, but he would demand a deep discount from their already reduced price, 50% off. NPC agreed and Suntech bought it for $400,000.

In November 2003, Suntech completed line #2 and it began full operation on December 18. By cobbling together used equipment, negotiating with suppliers Shi knew, and beginning to source from Chinese partners, Shi built the line for just over $1 million—an unheard-of cost of about 3 cents per watt of annual capacity. However, the initial yield—the share of cells that exit the process usable—was poor, only 80%. The remaining 20% were being lost from breaking during the screen-printing process. Shi called up NPC, explained the problem and told them, "don't let me down." NPC sent a team of engineers and within days had improved the yield from 80% to 98%. By the end of 2003, Suntech now had

40 MW of capacity running: the first 10 MW line and new 30 MW line. It had 300 employees in Wuxi. Despite this impressive growth, Suntech was still producing less than one percent of global PV.

Line #3: feed-in tariff

In early 2004, Shi again went to the Suntech board and told them he wanted to continue to expand. This time, Shi articulated his approach as the "low cost expansion strategy." They approved. He continued to expand on the cheap. He noticed that the US company Astropower had recently delisted from the NYSE. He knew that Astropower had two assets he was interested in. Suntech had bought a plasma enhanced chemical vapor deposition (PECVD) machine from German company Roth and Rau for Suntech's first line. Its nameplate referred to it as PECVD machine #3. Astropower had #1 and #2. Shi offered to buy them. One had been cannibalized for spare parts, but the other was operational and Astropower agreed to tell it to Suntech with a 50% discount. Shi had the machine sent to Hohenstein-Ernstthal, Germany for Roth and Rau to refurbish it and then on to Wuxi for Suntech line #3 via air freight. Line #3 began operations on August 18 and by the end of 2004, line #3 was producing 30 MW per year, for a Suntech total of 70 MW. This represented a seven times ramp up in two years but was still less than a fifth of the capacity of the industry leader, Sharp. Still, Suntech was ready for the industry explosion that was to come.

The German market took off in the second half of 2004. It grew by a factor of four in one year. Suntech had been waiting for the solar market to grow, but this was off the charts. Unlike the large established firms in Japan, Germany, and the US, Suntech was ready. The two large German suppliers, Q-Cells and Solarworld had distinct advantages: being close to their home markets, better access to suppliers, and existing scale. But they could not compete with Suntech on cost. Prices were falling and Q-Cells were barely able to sell their modules above their production costs. Meanwhile, Suntech made a net profit of $20 million in 2004.

Suntech really needed silicon wafers to make the modules it was producing, which they had been buying from Solarworld. In October 2004, Shi visited Solarworld CEO Frank Asbeck in Bonn. He went to Asbeck's home and discussed how Suntech could get more wafers in 2005. Asbeck joked, "Maybe this time next year, you will be supplying us!" while they walked in his garden. Silicon wafer supply was tight and German PV demand was growing. Knowing that Solarworld did not have the capacity to expand module production, Shi proposed that Solarworld continue to supply Suntech with 10 MW of wafers. If Suntech could get any more, Suntech would sell them back to Solarworld as modules which they could distribute in Germany. Asbeck found the proposal intriguing; they agreed, and soon had a deal. It was a good relationship to help serve the booming German market. Ironically, seven years later, Solarworld filed an anti-dumping grievance with the European Commission against Chinese solar companies. Shi called him and asked, "Frank, what are you doing?" Shi acknowledged that the world had changed and "Frank needed to save his own life."

This relationship reveals that Suntech never wanted to make its own wafers. Shi's philosophy was one of comparative advantage—each company should specialize in their own niche of expertise. But Suntech became the first solar company to develop downstream PV projects. It nearly entered the development business in 2006 by considering an acquisition of a Kingston, NY PV developer. Shi was eventually dissuaded from closing the deal by the argument: "if you start competing with your customers, who will want to buy from you?" Shi still felt that doing something to expand demand for PV would be beneficial. He would return to that strategy two years later with the fateful acquisition of the Global Solar Fund (GSF).

Shi acknowledges that new problems emerge at scale, even as the process improves with experience. Capital costs become substantial, but more than that, talent and training become central concerns. In the early days, Shi would train employees in the evenings three times a week. As the company got bigger that tutoring style approach was no longer feasible. The company had to develop procedures and training in a more formal way. That was difficult and did not align with the company's bootstrap beginnings.

> Not everyone can start a company. How you get a team together is key. I always miss those days. We would meet on the weekends to talk about what was missing on the production line and why. People would fight over these discussions. But they would hug each other after. Company culture was very important. It had a calming effect despite the frenzied pace. –Zhengrong Shi

Shi says that over the past 40 years, "speed" has been one of the key factors in China's success. There are so many entrepreneurs who are quick to spot an opportunity and take advantage of it. Doing a quick and dirty low-cost version has been a hallmark of the beginnings of many firms in China that later grew enormously.

2006 Renewable Energy Law

In 2003, the National Development and Reform Commission (NDRC) created a new renewable energy division. The central government had decided to do more with renewables. It began by enhancing the village electrification program with a RMB3 billion from 2003 to 2005. Many firms benefited from that period, including Yingli and Trina. However, the experience with rural electrification made it clear that PV faced barriers to deployment, particularly in that PV adopters were not always able to connect their system to the power grid. They needed a new law to remedy this. In 2003, with assistance by the German government as well as Denmark and the US, the Chinese designed a renewables law. The plans for the law were announced in 2004 at the international renewable energy conference in Bonn. In January 2006 it went into force. China now had a national renewable energy law. In its early years it mainly supported wind power, since PV was so much more expensive. But it had an important effect on perceptions of renewables, creating an expectation that China's domestic market PV would, at some point, emerge and could become very large. That was a crucial signal to foreign investors who were becoming increasingly interested in China's solar firms.

Initial public offering

Other Chinese companies had begun to see what Suntech was doing and how they were doing it but lacked Suntech's international certification and international distribution networks. However, Suntech lacked inputs: it didn't have enough silicon to create the silicon wafers it needed, nor did it have ingot slicing capability. Their method was to purchase ingots from their Chinese suppliers, send them to Japan or Norway to be sliced and shipped back as wafers. That process was expensive. Shi intervened. In early 2004, he brought the first multi wire saw to China from Meyer-Burger in Switzerland because he knew their CEO, Peter Pauling. China was now producing silicon wafers at scale. Suntech began sourcing from China other inputs that had previously only been available expensively from abroad: glass (to cover the cells) and aluminum paste (for electrical conductors). Suntech helped these companies establish themselves in China, not through money, which at the time Suntech was short on, but through international connections and expertise, which it had in abundance. Suntech had established key parts of the supply chain in China.

Suntech's financial constraints changed considerably in 2004. US venture capitalists became interested in both solar and China more generally. Suntech attracted $100 million of Western venture capital investment, led by Goldman Sachs, DragonTech Ventures, and Actis Capital. Around the same time, Suntech management organized a $100 million buyout of the original investors. The original appliance manufacturers in Wuxi gained a return on their original $1 million investments by a factor of 16 in less than four years. Suntech was now clear of state ownership, a requirement for US equity markets. In December 2005 Suntech raised $400 million in an initial public offering. It demonstrated its new clout even more when it bought a Japanese company, MSK, which had developed an assembly technology as a supplier to Sharp. Up until then it was rare for the Japanese to see investment from China rather than the other way around. This was a new channel through which China acquired know-how from abroad. Suntech was now influential in China. Up until this point, the Chinese Central government's focus in renewables was wind, which was far more affordable than solar. Suntech began attracting government support and financial support. Its success on Wall Street showed what could be done in solar in China.

A Chinese solar industry

Suntech's highly visible success created a replicable path. In China, such opportunities are rarely missed and many other followed.

Early entrants

Followers emerged to emulate Suntech's success. The first followers had actually started prior to Suntech, serving the small but growing market in rural electrification under the Brightness Program. Using Suntech's model, they eventually scaled up to levels well above those seen before.

Like Suntech, Yingli was an entrepreneurial venture, not an arm of a Chinese state-owned enterprise (SOE) or a large conglomerate like its competitors in Japan. Liansheng Miao had started several companies in other areas, such as water purification and cosmetics. He founded Yingli in 1997 in Baoding with $600,000 of from personal funding. He had seen solar powered neon lighting in Japan and started importing fixtures to China in 1993. As an entrepreneur and businessman, he became interested in starting a solar company. While this entrepreneurial aspect provided nimbleness it also involved obstacles. For example, in 1998 Yingli was denied a $3 million loan from China Development Bank because only SOE's were eligible for such financing (Dong et al., 2015). To get around this, the government of Baoding High-tech Zone arranged for one of its firms to purchase a majority of Yingli. Yingli was then able to hire a CTO from the Beijing Solar Research Institute, Yuting Wang, and began to re-orient the company to target the PV export market.

From there, Yingli followed Suntech's production scale up trajectory. Unlike Suntech, which cobbled together a line from used equipment, Yingli bought a turnkey module production line in 2001. It set up a 3 MW cell line in 2003 with used equipment from Amtech, GTAT, and others and utilized technicians from those companies to make it operational (Zhang and Gallagher, 2016). Yingli sold its first modules to Germany in 2004 accompanied by heavy advertisement in Germany and Spain. It hired people from the Chinese semi-conductor industry, enabling them to double production almost every year. In June 2007, 18 months after Suntech's IPO, Yingli went public on the New York Stock Exchange and raised $319 million.

Similarly, Trina was started by entrepreneur Gao Jifan, who had no background in solar, with credentials including a Master's degree in chemistry and production experience in household detergent (Zhang and White, 2016). Gao founded Trina in late 1997 in Changzhou, initially to produce an energy efficient exterior house wall and then targeting off-grid PV systems, the first of which was installed in 2000. Trina became the first non-SOE to become a supplier to the Brightness Program, which raised Trina's profile and enabled it to attract engineers from the semi-conductor industry. Once the German market took off, Trina scaled up quickly, producing its first module with a 6 MW line in late 2004 using wafers it bought from Sharp. Trina scaled up more slowly than Yingli and Suntech, only surpassing 100 MW of capacity in 2007, the year after it went public in the US. From there they invested heavily, with $80 million of new plant construction in 2007 including 20 Chinese ingot pulling machines, 20 Swiss wire saws, 4 German diffusion furnaces, and 6 Chinese automatic laminators.

Xiaohua Qu started Canadian Solar a year after Suntech. Like Shi, Qu was a PV scientist with a technical orientation from the beginning of his career. Qu had been trained in material science from the University of Toronto and worked as a researcher at Ontario Power Generation Corporation. He also developed business development skills working at ATS that transformed into his position as Vice President for Asia at Photowatt, a PV company bought out by ATS (Zhang and White, 2016). While at Photowatt, Qu developed a business plan for a solar battery system for vehicles and obtained an order from Volkswagen for it. He raised $400,000 in angel investment and began manufacturing it in Changsu in late 2001. The company set up production of modules in Suzhou in 2004, surpassing 100 MW in the same year that Trina did. It listed on the NASDAQ in November 2006 that it raised $108 million. In 2007 it began manufacturing cells.

Whereas Suntech and Canadian Solar were started by a team with more than a decade of experience in producing world-class solar cells, Yingli and Trina were not led by PV experts, although they later hired Australian-trained engineers as Chief Technology Officers. They wisely took a more incremental approach. They each started production with the least technically challenging step of production, creating solar modules by wiring together cells and encapsulating them in glass, metal, and plastic. They imported cells, mainly from Germany. These firms focused on buying cells, producing modules, and exporting them as panels to Germany. Activity in this period was not capital intensive; Chinese labor cost advantage was a key source of comparative advantage in the early days of the industry. Later they upgraded into more complex activities, such as Trina's expansion into poly-silicon production in 2005, then into wafers in 2006, and cells in 2008 (Zhang and Gallagher, 2016).

The "fast follower" strategy allowed for some technology leapfrogging as better equipment became available to companies like Trina and Yingli who scaled up later. These firms continued to expand their technical capabilities, both launching R&D programs by 2007 with engineers trained at the University of New South Wales in Australia. Yingli launched an R&D program in 2007 and Trina did so before that. For Trina, the R&D was increasingly tied to production. It set up a small pilot line, its "Golden line," adjacent to its main production line, to test out improvements in production (Ball et al., 2017). By 2014, Trina had developed a world record efficiency for a multi-crystalline solar cell.

Silicon supply

In 2006 a new challenge emerged. The price of purified silicon, the main material input for making PV wafers, had been rising since 2004. In 2006, Suntech agreed to buy $6 billion of wafers from Monsanto Electronic Materials Company (MEMC) in the US over 10 years. In late 2007 and early 2008 prices spiked to nearly $500 per kg, a ten-fold increase in four years (Figure 7.2). It coincided with the dramatic increase in demand for PV systems in Germany. Getting access to silicon became a serious issue in 2008 (Pillai, 2015). Nitol, based in Irkutsk Siberia,

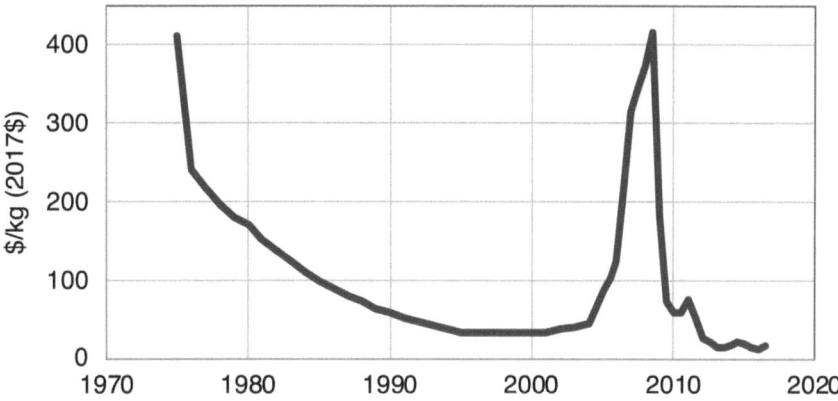

FIGURE 7.2 Price of purified silicon.

became a key supplier. In 1998, Nitol founder Dmitry Kotenko had bought a 70-year-old chemical plant that produced trichlorosilane and reoriented the business to focus on poly-silicon for the solar industry, for which trichlorosilane is a precursor chemical. With funding from Russian banks and $75 million from the World Bank, Nitol invested $700 million in the plant at Novocheboksarsk, buying $50 million worth of poly-silicon reactors from GT Solar in the US as well as machines from Oerlikon and Renova. To secure supplies for module production, Suntech bought a $100 million minority stake in Nitol in 2008, including a 5-year procurement contract. Trina also participated in securing supplies through Nitol by signing a 5-year supply contract in 2009 200 MW worth of poly-silicon.

Unsurprisingly, Chinese companies became interested in poly-silicon production. Since making poly-silicon is relatively simple once one has equipment to do so, some PV companies decided to produce their own poly-silicon. For example, Yingli used furnaces from GTAT to build their own poly-silicon facility that came on line in 2010 (Zhang and Gallagher, 2016). But the equipment is expensive and extremely energy intensive, involving furnaces heating a silicon to containing gas above 1100° Celsius. For many years, a small group of companies produced silicon: Hemlock and MEMC in the US; Wacker in Germany; REC in Norway; and Yokuyama, Mitsubishi, and Sumitomo in Japan. Unlike other equipment needed for PV production, these suppliers had been mainly unwilling to sell equipment to the Chinese. Once prices rose, around 40 companies were started. Applied Materials had boom years selling furnaces to companies there were setting up silicon production. These startups paid dearly for furnaces and fabrication lines. But with prices above $400 per kg, companies could afford high levels of upfront capital investment.

Peng Xiaofeng founded LDK in 2005 and went directly into wafer manufacturing. It started its first 200 MW line on May 2, 2006, having purchased equipment from GT Advanced Technologies of the US. LDK went public in New York on June 7, 2007 and by 2008 it was the largest wafer manufacturer in the world. Seeing the same rising silicon prices that others saw, in 2009 LDK started manufacturing high purity silicon.

Much like an oil boom creates the conditions which lead to its bust, the silicon price spike in 2007–08 did what prices are supposed to do. They signaled opportunity and stimulated entry of new production into the market. GCL, LDK, OCI, and Renesola all entered silicon production at this time. Prices dropped by a factor of ten in 2008–09 coinciding with the global financial crisis. Yingli and Suntech each lost hundreds of millions of dollars on their commitments to high priced silicon. But they stayed in the game of PV production and prices returned to their early 2000s levels and have stayed below $30 per kg since 2012.

A supply chain

At the time of Suntech's founding in 2000, China had none of the core necessary elements of the PV industry. It had no substantial market, no means for exporting, a small talent pool, no precedents for investing, and none of the important participants in the manufacturing supply chain. By 2005, Suntech had established all of these elements, allowing others to follow. Shi credits two important developments to Suntech's success. First, he relied on his extensive international contacts built up over nearly two decades on the cutting edge of PV research at UNSW. Second, the shortage of silicon made him realize that he eventually would need a local supply chain. It needed to be reliable and inexpensive. Suntech helped develop that supply chain for itself, and eventually for other entrants, to make use of. Suntech established a supply chain cluster in the Yangtze River Delta in which many other PV companies located, including Trina and Canadian Solar. The success of the solar sector entrained other companies into it such as glass producer Flat Glass, which got into solar in 2006 and is now the world's largest producer of glass for solar panels.

Suntech trained their local staff. They made use of a "simulated production facility," conceived and developed by Professor Stuart Wenham at UNSW, with exercises to teach production managers to tweak processes, with Wenham traveling to Wuxi for 1–2-week periods to train staff. This practice was based on a class at UNSW in which one's grade for the class was based on the yield one could achieve in the simulator. Suntech got many engineers up to speed quickly, many of whom were headhunted by the other firms that emerged and brought their knowledge to them. Wenham ended up becoming CTO of Suntech, prior to its listing on the New York stock exchange, and was very effective in the preceding "road shows" in generating investor interest.

Suntech exceeded a billion dollars in sales in 2007 and reached a production capacity of a gigawatt in 2008, making it the largest producer in the world. Despite the high silicon prices, it was able to produce modules below $3 per Watt. In 2009 a second public offering raised $250 million and set a new world record for multi-crystalline module efficiency. It surpassed 2 GW of capacity in 2011 and, once silicon prices fell, the cost of making modules plummeted. Suntech was producing modules for less than $1.50 per Watt in 2011.

Having created their own supply chain that enabled growth, Suntech shifted focus to grow the other end of the market, the "downstream" end, by providing finance for solar projects. It invested in and eventually owned a majority stake in an Italian company called GSF (Global Solar Fund) which did project development in Europe. Suntech completed the first utility scale solar project in China in 2009.

Later entrants

Another set of Chinese solar companies emerged which were founded after the German market's dramatic emergence in 2004. Solarfun and Sunergy were both founded that August and both recruited management from Suntech and those trained in Australia. The following year, JA Solar, LDK, and ReneSola were also founded. All five went public by mid-2007. Another, Jinko launched in 2006 and went public in 2010. These six firms raised $4 billion from investors in US markets, from the initial public offerings as well as follow on offerings of their shares. Many of these firms also received substantial funding from the local governments in which they were located (Xia, 2013). They took full advantage of the business model that Suntech established, the supply chains that emerged, and the market that continued to grow—mainly in Germany, but also in Spain, Italy, and California.

Whereas the early Chinese firms typically purchased individual pieces of equipment and connected them into a line themselves, some in this latter group of entrants were able to purchase sets of machines that were already integrated together by the suppliers. These so called "turnkey" production lines could handle the entire cell to module process. Centrotherm had started selling turnkey lines to Taiwanese companies such as Gintech, Neosolar Power, and Solartech in 2003–04. They became widely available in China in 2007 and 2008, with Applied Materials as an important supplier of them (Dewald and Fromhold-Eisebith, 2015). Many in this group of firms scaled up from initial funding to 100 MW production capacity in less than two years. That same process took Suntech five years. Expectations of continuously growing markets fueled these investments and made Western investors eager for shares in their profits, but problems soon emerged in those expectations.

Collapse

The first jolt to growing markets came in the summer of 2008, and later became known as the global financial crisis of 2008–09. PV is a highly capital-intensive energy source, in which almost all of its costs are borne before any electricity is produced. So, PV adoption was highly sensitive to the deteriorating access to credit and rising commercial lending rates. An early consequence was the reduction in Spain's 2007 FiT which stimulated over 3 GW of installations in 2008. The lack of economic activity and increasing deficits put pressure on Spain to cut the rates it offered for PV and cap installations at 500 MW in 2009.

The German market also faced increasing pressure to reform. The Green Party fell out of power in 2005 and thereafter electric utilities and energy users influenced the government to reduce the FiT rates. In 2011, plans were put in place to slash the German FiT. Instead of continuing to grow at over 40% per year, the German market grew at less than 1% in 2011 and in 2012, and then shrank after 2012. During this time, Suntech's revenues declined by nearly 50%. It had a negative gross margin, meaning it was losing money on each module it sold. This lack of cash flow put pressure on its finances. But Suntech's troubles were only beginning. In the wake of the global financial crisis, Suntech had guaranteed a $683 million line of credit to one of its largest subsidiaries, GSF to help expand their international investment portfolio. In mid-2012, it emerged that GSF had forged half a billion euros in government bonds, which it had been using as collateral to secure Suntech's credit line. Suntech, as majority owner of GSF with an 80% stake in the company, now had a major debt problem, as well as a credibility problem. Its share price dove and in 2013, Suntech defaulted on its debt leading the Suntech board to fire Shi before bankruptcy proceedings began.

Despite its demise, Suntech's legacy survived. Trina, Yingli, and Jinko are now producing at 5–7 GW capacities. Nearly the entire PV supply chain is located in China, even the equipment. Other firms took the Suntech model and in a bigger way than Suntech ever did, many with more modern technology and equipment. Capacity utilization declined between 2012 and 2017, yet expectations of future demand remain high as new markets emerge.

The Suntech story involves somewhat of a paradox in that Shi emerged from UNSW, the most elite PV research group in the world, famous for making the most efficient cells in the world. And yet Shi made his greatest impact in scaling up production of those technologies in the most low-cost, quick, and competitive race there could be. Amazingly he worked it at both ends. Shi used his elite technology training to convince the skeptical Germans to buy from him. He used his scrappy resourcefulness to scale up quickly with limited access to capital and, eventually, reduce the cost of production by a factor of three. China's start in PV needed both of these complementary attributes.

Home market

Though Suntech went bankrupt in 2013, its competitors were saved by a somewhat unexpected development—a domestic Chinese market. As one PV executive put it in 2008, just before the global financial crisis,

> Today looks sunny. But tomorrow might be dark skies. And even the day after tomorrow might also be dark skies. But we know after that that it will be sunny and we are going to be ready for that.

Why the optimism? First, PV's success in Western equity markets in the 2005–07 period raised PVs prominence in the eyes of the central government. It began to signal its support for PV as an emerging strategic industry. On one hand, this support was part of a broader effort by the central government to stimulate domestic demand to compensate for the drop in exports due to the global financial crisis. It launched a half trillion-dollar economic stimulus package in 2009–10. More specifically the "Decision of the State Council on Accelerating the Fostering and Development of Strategic Emerging Industries" identified "new energy" as one of the seven industries to be championed. PV was part of this new energy and the Chinese Development Bank was made responsible for supporting PV, providing lines of credit worth $30 billion to PV manufacturers (Grau et al., 2012).

Second, in 2009, the Chinese central government launched three programs that created a domestic Chinese PV market. The Ministry of Finance and the Ministry of Housing and Urban Rural Development announced The Rooftop Subsidy Program, which provided a $2.50 per Watt up-front rebate for residential solar installations. The Golden Sun Demonstration Project provided a 50% up-front subsidy for grid systems between 2009 and 2013. The government began soliciting bids for utility scale PV projects in 2009 through the "Large-scale PV Power Station Concession Bidding" program.

Third, in 2011 China initiated its own FiT, motivated in part by the financial crisis and the ensuing reduced demand abroad. In 2011, the US imposed tariffs on Chinese PV and the EU soon followed (Kolk and Curran, 2015; Deutch and Steinfeld, 2015). Further, PV had become quite inexpensive since subsidizing deployment was getting cheaper and easier. The "Notice on Perfecting Feed-in Tariff Policy of Solar PV Power Generation" provided a 17c per kWh rate for 3–5 years (Dong et al., 2015).

These policies succeeded in creating a market of over 10 GW by 2013 making China the world's largest PV market. In 2017 over half of the new solar capacity in the world, 50 GW, was installed in China. To put that in perspective, that level is seven times larger than the German market was at its peak in 2012. Curtailment of PV production became a significant issue. Importantly, the inevitability of more renewables seemed to be acknowledged when the national transmission company, State Grid, previously no friend of intermittent sources, began to work with provinces to enable cooperation and modernization of grid operations. Again, wind was a helpful precursor technology in that the experience with managing intermittent wind helped with solar later (Qi Ye, 2018). In 2016, the central government targeted 110 GW of installed capacity by 2020, which was surpassed by 2017. Continuation of recent 50% per year growth rates would get China to 110 GW per year by 2019. Costs fell dramatically during this period (Figure 7.3). The low cost solar that emerged at the end of this period was China's own gift to the world.

Upgrading

As the home market grew, technology upgrading occurred throughout the supply chain. This started to emerge in the IPO period 2005–08, as firms began to make their own poly-silicon as well as system components like inverters (Zhang and

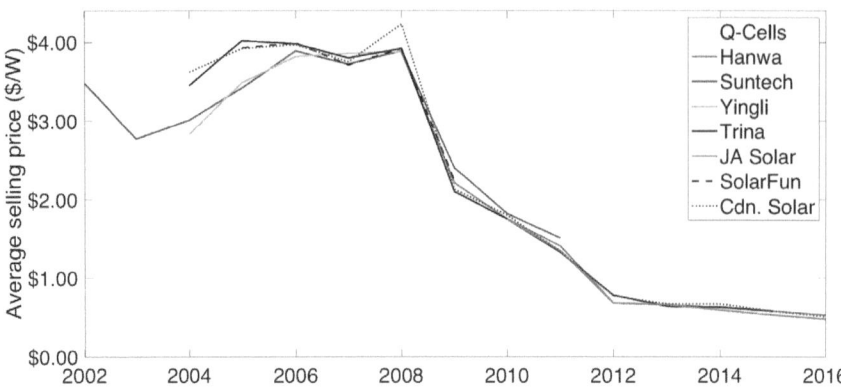

FIGURE 7.3 Average selling price of modules.

Gallagher, 2016). Firms' R&D spending began to rise substantially after that, as they established new world record efficiencies and used process improvements to reduce manufacturing costs. Starting around 2007, the Chinese worked hard to make their own manufacturing equipment rather than buying it abroad. Chinese firms now manufacture diffusion furnaces, plasma etching machines, cleaning-spinning machines, and drying ovens. Key emerging challenges are to reproduce the most difficult technical challenges such as making silver paste, ethylene vinyl acetate (EVA), wire saws, automatic welding machines, screen printing, chemical vapor deposition, and voltage testing equipment (Ball et al., 2017). Collaboration networks among Chinese suppliers are becoming increasingly robust (Liang and Liu, 2018) although they still remain far from optimal (Zou et al., 2017). An important effect of China's interest in upgrading and demonstrating competence in doing has been that it has cut into the margins of Western equipment suppliers who now must price aggressively to discourage China's entry into equipment supply. Chinese producers benefit from this implied threat and its perceived inevitability over the longer term.

China-based possibilities seem to be growing. In January 2015, Liu Lifeng was a 21-year-old undergraduate in electrical engineering at Tsinghua University. Using the high efficiency PV technology he was working on, he submitted an entry to a business plan competition and won. The prize was access to Tsinghua's well-connected and wealthy alumni network. In April 2015 he founded Beijing Sunlectric Technological Company. Tsinghua alumni investors enabled Liu to purchase equipment and set up a factory. The company owns 51 patents and claims to have the world's highest efficiency for a commercial crystalline solar panel at 22.1% (although US-based SunPower now offers a 22.7% efficient panel).

Liu commutes via a 2-hour flight between his factory in Chengdu and Beijing where he is still taking classes for his undergraduate degree in electrical engineering. He worries he is spending too much time at the factory and not enough in school. He is considering entering a PhD program. Liu doesn't speak much English. He doesn't need to. The machines and robots in his factory all come from China. The silicon material comes from China. Sunlectric does installations as well. They install more than they can make. Their customers are all in China. That is completely different than the early 2000s when Suntech started serious PV manufacturing in China when manufacturing technology came from the West, know-how came from the West (particularly Australia), and products were sold to the West (Hopkins and Li, 2016).

Liu collaborated with the local equipment manufacturers to modify the machines so that they worked for his high-performance cells. He tried partnering with a few companies but some were unable to make machines capable of producing cells at his needed specifications. Eventually he found a reliable machine maker and he is now able to get yields of 99% for his 22% efficient cells. Consider that this would not have been possible even five years earlier when the machines came from Europe and customizing them to extreme efficiencies would have been unthinkable. His factory is almost entirely automated—robots abound and few humans. Liu is cautious about showing it.

As of early 2018, Liu's factory was capable of making 100 MW per year, but he needs it to be bigger. He can make panels for $0.40 per Watt, but he needs to be cheaper. The limiting factor is accessing financing, as 20% of the firm's costs are in servicing debt. The largest PV firms now are producing over 6 GW per year, dwarfing Sunlectric. These massive plants can produce a cell for a little over $0.20 per Watt. Still, the fact that Liu is able to produce at costs not far away from the giants suggests that China's advantage is not just massive scale. Its advantage now rests on new technology and processes, which still have ample opportunity to realize cost reductions through their own as-yet-untapped scale economies. Above all, China's advantages depend on an intensely entrepreneurial and competitive culture that is making full use of scale, technology, and know-how.

References

Ball, J., Reicher, D., Sun, X. & Pollock, C. 2017. *The New Solar System: China's Evolving Solar Industry and Its Implications for Competitive Solar Power in the United States and the World*, Steyer-Taylor Center for Energy Policy and Finance.

Binz, C., Tang, T. & Huenteler, J. 2017. Spatial lifecycles of cleantech industries – The global development history of solar photovoltaics. *Energy Policy*, 101, 386–402.

Deutch, J. & Steinfeld, E. 2015. *A Duel in the Sun: The Solar Photovoltaics Technology Conflict between China and the United States*, Cambridge, MA, MIT.

Dewald, U. & Fromhold-Eisebith, M. 2015. Trajectories of sustainability transitions in scale-transcending innovation systems: The case of photovoltaics. *Environmental Innovation and Societal Transitions*, 17, 110–125.

Dong, W., Qi, Y. & Spratt, S. 2015. *The Political Economy of Low-Carbon Investment: The Role of Coalitions and Alignments of Interest in the Green Transformation in China*. IDS Evidence Report160, Brighton, IDS.

Dunford, M., Lee, K. H., Liu, W. & Yeung, G. 2013. Geographical interdependence, international trade and economic dynamics: The Chinese and German solar energy industries. *European Urban and Regional Studies*, 20, 14–36.

Grau, T., Huo, M. & Neuhoff, K. 2012. Survey of photovoltaic industry and policy in Germany and China. *Energy Policy*, 51, 20–37.

Green, M. 2016. Revisiting the history books. *PV Magazine*, 96–101.

Green, M. A. 1982. *Solar Cells: Operating Principles, Technology, and System Applications*, Englewood Cliffs, NJ, Prentice Hall.

Hejun, L. 2015. *China's New Energy Revolution*, New York, McGraw-Hill Education.

Hopkins, M. & Li, Y. 2016. The rise of the Chinese solar photovoltaic industry: Firms, governments, and global competition. In: Zhou, Y., Lazonick, W. & Sun, Y. (eds.) *China as an Innovation Nation*, Oxford, Oxford University Press.

Kolk, A. & Curran, L. 2015. Contesting a place in the sun: On ideologies in foreign markets and liabilities of origin. *Journal of Business Ethics*, 142, 697–717.

Liang, X. & Liu, A. M. M. 2018. The evolution of government sponsored collaboration network and its impact on innovation: A bibliometric analysis in the Chinese solar PV sector. *Research Policy*, 47, 1295–1308.

Lou, J. 2008. The reform of intergovernmental fiscal relations in China: Lessons learned. In: Wang, S. and Lou, Y. (eds.), *Public Finance in China: Reform and Growth for a Harmonious Society*, Washington, D, The World Bank.

Pillai, U. 2015. Drivers of cost reduction in solar photovoltaics. *Energy Economics*, 50, 286–293.

Qi Ye, L. J. A. Z. M. 2018. *Wind Curtailment in China and Lessons from the United States*, Beijing, Brookings-Tsinhua Center for Public Policy.

Quitzow, R. 2015. Dynamics of a policy-driven market: The co-evolution of technological innovation systems for solar photovoltaics in China and Germany. *Environmental Innovation and Societal Transitions*, 17, 126–148.

Xia, Y. 2013. *The Role and Incentives of Chinese Local Governments in Solar PV Overinvestment*. MA thesis, University of Texas, Austin.

Zhang, F. & Gallagher, K. S. 2016. Innovation and technology transfer through global value chains: Evidence from China's PV industry. *Energy Policy*, 94, 191–203.

Zhang, W. & White, S. 2016. Overcoming the liability of newness: Entrepreneurial action and the emergence of China's private solar photovoltaic firms. *Research Policy*, 45, 604–617.

Zhao, Z. Y., Zuo, J., Feng, T. T. & Zillante, G. 2011. International cooperation on renewable energy development in China – a critical analysis. *Renewable Energy*, 36, 1105–1110.

Zhi, Q., Sun, H., Li, Y., Xu, Y. & Su, J. 2014. China's solar photovoltaic policy: An analysis based on policy instruments. *Applied Energy*, 129, 308–319.

Zou, H. Y., Du, H. B., Ren, J. Z., Sovacool, B. K., Zhang, Y. J. & Mao, G. Z. 2017. Market dynamics, innovation, and transition in China's solar photovoltaic (PV) industry: A critical review. *Renewable & Sustainable Energy Reviews*, 69, 197–206.

8

LOCAL LEARNING

While the extent of cost reductions in PV hardware has been dramatic, costs of installing solar have also fallen (Ardani and Margolis, 2015). Using industry jargon, these "balance of system" or "soft costs" involve, inter alia, labor in installation, insurance, marketing, and obtaining permits (Friedman et al., 2013). Because hardware has become so inexpensive, soft costs represent approximately two-thirds of the costs of an installed residential PV system in the US (Barbose et al., 2017). These costs are distinct from hardware costs in that they are almost completely locally driven (Karakaya et al., 2016), whereas PV modules are globally traded. PV is now so inexpensive because of innovation happening at the local level—both in adoption and reductions in soft costs.

First, PV adoption decisions have been local. In contrast to almost all other energy supply technologies, for which large companies make decisions, households have by far been the most important adopters of PV systems. The small scale and modular design of a PV system—a typical panel is 200 W and about 20 are needed for a household—put decision making in the hands of customer. This makes PV more like a typical consumer product. It has many of the characteristics, and similar costs, of a consumer durable product like hot water heaters, furnaces, and to some extent automobiles. As a result, PV shares some of the adoption drivers with those. A cost of $10,000 or greater makes financing important. The one-time nature of the purchase requires information and imposes search costs for adoption. Unlike furnaces and hot water heaters, but similarly to cars, installed PV systems can be a form of conspicuous consumption, an expression of identity. Neighbors who have installed solar are important drivers of adoption in that they provide trusted sources of information. Finally, households have often been willing to pay more for PV than the value of the electricity they would avoid having to purchase. These consumers have thus provided an essential niche market for PV.

Second, this local adoption activity also led installers to scale up and develop experience in the marketing, installation, and financing of PV systems. With its own learning curve, local activity has generated substantial reductions in the prices paid for installed systems. While PV modules are fully globalized with a far-flung supply chain and easily transported end products, the rest of the cost of a PV system is almost entirely local. Moreover, as the prices of modules have fallen, the non-module, or "soft" costs have made up an increasingly large proportion of the total costs. In US residential installations today, those soft costs are close to 80% of the total costs (Figure 8.1). But those costs are also falling. Improvements are being made at the local level in which marketers, financers, installers, and others in the market are identifying ways to improve performance and reduce costs. Those actors are learning from each other. Subsidies raise prices in the near term but reduce them in the longer term by increasing demand and creating opportunities for learning by doing and economies of scale.

Third, solar's adoption has prompted new challenges, which also are rooted in local dynamics. As PV became widely adopted, concerns have risen regarding grid stability, value deflation, electric utility business models, related political economy issues, and financing. The latter is an increasingly prevalent issue, as PV has large potential in developing countries where credit is often scarce.

For both adopters and installers, peer effects are occurring at the local level. Potential adopters are learning about PV from their neighbors, while installers are learning from the experience of other installers. The local aspects are important because they have sustained interest and political support from a wide variety of constituencies over a substantial period of time. The local activity, especially these interactions among peers, has also been expressed in political advocacy which has effectively sustained support for PV despite the centrifugal forces of globalization and the boom and bust nature of the industry. In some places, local stakes in the outcomes of PV related policy provide political support that made support for PV

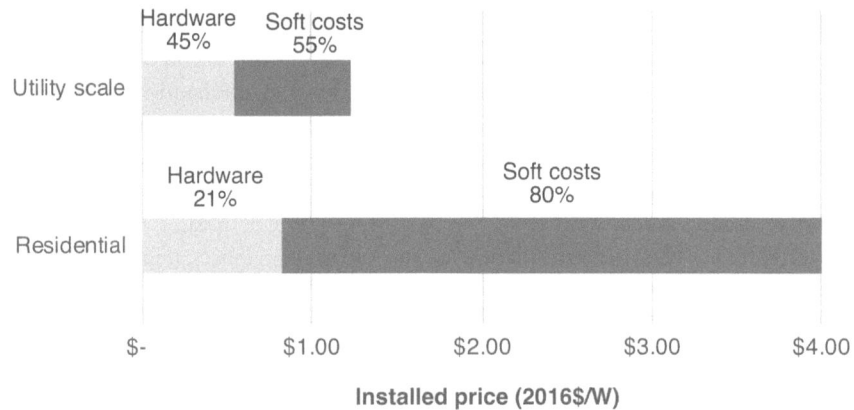

FIGURE 8.1 Share of costs for hardware (globally sourced) and soft costs (local) in the US in 2016.

robust to the forces of globalization which tended to move the locus of production and technology hardware development.

Local adoption processes

Local factors have substantially affected the adoption of PV. Only since 2015 have utility scale projects become a substantial part of the global PV market, and even since then residential PV continues to account for most of total global installations. Local policies, such as in Aachen and Hummelsberg created the first residential markets. Beyond policies, neighbors have and continue to play an important role in adoption. This local adoption process makes PV distinct from nearly every other energy supply technology, which tend to be large and adopted by companies or public utilities. Rather than a few, or a few hundred, large decisions in those cases, PV adoption has involved millions of small decisions to adopt PV—and potentially many times that, explicit decisions not to adopt PV. This local, small, and massively repeated adoption process has involved distinct characteristics, which in turn have led to distinct cost reducing activities in soft costs.

Since its early days, a combination of financial and non-financial incentives has driven adoption of residential PV. Over time, as PV costs have fallen, adoption has shifted from early adopters with high willingness to pay to mass market adopters who more closely assessed the financial payoffs of installing PV (Moezzi et al., 2017). Financial incentives have thus played a more important role for these more recent and much more numerous adopters. Starting with the first FiT in Aachen in 1994, an array of policies has been implemented to create financial incentives for residential PV: rebates for up front capital costs, tax credits, guaranteed power purchase agreements (FiTs), solar renewable energy credits, and renewables obligations. The stringency of these policies has been generally declining over time as the costs of PV without subsidies have declined. In the US however, the most important financial incentives have been the value of savings from not purchasing grid electricity (Nemet et al., 2017a). This value depends on the electric rates that customers would face without PV, whether they can sell excess power back to the grid (net metering), and the prices customers are paid for the electricity they produce. Setting the levels of these parameters to create the optimal amount of PV adoption has proven challenging (Benthem et al., 2008). In some cases, such as the ISO4 in California in the 1980s, levels were never high enough to create a financial incentive for substantial adoption of PV. In other cases, such as in Japan in the early 1990s, a small incentive program generated more interest than expected, giving policymakers some sense of the price elasticity of demand for PV. In other cases, such as the German FiT from 2004 to 2010, financial incentives were perhaps excessively high, leading to shortages, price spikes, short-term focus, and suboptimal use of ratepayer funds for the program. As the price of PV has declined, these financial incentives have needed to be much less stringent to stimulate adoption. However, as they have pushed adoption to consumers with lower willingness to pay for PV, they have become important to the adoption decision. Even as their importance has diminished, non-financial incentives still continue to drive adoption in combination with financial incentives (Sommerfeld et al., 2017).

PV is distinct in that its attributes lead to a demand for PV, not just tolerance of it. To many PV is appealing, something many people want if they can find a way to pay for it. While most consumer products work this way, energy technologies do not. One usually speaks of energy technologies as being "accepted" rather than demanded (Rand and Hoen, 2017). Much of the concern about other low-carbon technologies such as nuclear power, carbon capture, and solar radiation management focuses on how to deal with problems in public acceptance (Krause et al., 2014). PV works in the opposite direction—people want it, rather than protesting to avoid it.

That people find PV appealing also gives it a distinct epidemic-style adoption pattern. An array of research points to the importance of "peer effects" in influencing PV adoption (Bollinger and Gillingham, 2012). Even after controlling for socio-economic characteristics, people are more likely to adopt PV if a neighbor has already done so (Rai and Robinson, 2013). In recent work colleagues and I found a similar result through examination of the hypothesis that neighbors provide a low-cost channel for high quality information about installers, options, and pricing (Nemet et al., 2016). A challenge throughout these studies is to disentangle the effect of neighbors' proximity from the effects of neighbors' similar demographics. Since PV tends to be more likely to be adopted by those with high incomes, high education, and left political leanings—and since those attributes tend to be clustered themselves—the challenge with these studies is to find an additional effect from a neighbor's installation. Research results show this additional effect exists. The result of these neighbor effects is that adoption appears to proceed in clusters where one household's adoption helps beget the adoption of neighbors. The interaction of peers, social norms, and policy design is an emerging policy area (Nyborg et al., 2016), one where solar adoption may provide a model.

Residential PV adopters value solar not just for the financial value it generates, but for a variety of other attributes: its displacement of fossil generation to avoid air pollution and climate change, its local employment benefits, the communication of green identity, and its enhancement of self-sufficiency through decreased dependence on the grid. One is struck by the size of this cohort of PV adopters in several countries who value the non-financial attributes of solar, including Japan (Nomura and Akai, 2004), India (Urpelainen and Yoon, 2015), and South Korea (Lee and Heo, 2016). Adopters of PV show a willingness-to-pay (WTP) of 2–4 cents per kWh above traditional energy sources (Sundt and Rehdanz, 2015), a premium of about 30%. Colleagues and I have found that this above-market WTP provides a partial explanation for the persistent bias that models of future energy systems, which aim to minimize energy costs, have shown in underestimating PV adoption (Creutzig et al., 2017).

Niche markets like satellites, remote telecom repeater stations, and navigational aids have been critical to the adoption of PV. Consumers with high willingness to pay for PV have been another crucial niche market, one that has been more substantial to the success of the industry, and thus more important to the sequential niche market progression toward competing with other technologies for large-scale power generation. This local adoption, and the clustering that it has engendered, has led installers to develop ways to meet this demand, to scale up, and to learn from the process of procuring, selling, and installing PV systems. That process has led to its own set of important cost reductions.

Local cost reductions

Even though much of the attention has been on module costs, soft costs have also fallen—by 48% between 2010 and 2017 (Fu et al., 2017). Soft cost reductions also have been locally driven—by installers improving their processes, learning from each other, and to some extent becoming more competitive with each other in part due to potential customers having better access to information. In the US for example, 49% of the cost reductions between 2000 and 2017 were on soft costs (Figure 8.2). This result is additionally impressive in that the US still has higher soft costs than other countries such as Germany (Seel et al., 2014). There is thus still plenty of room for soft costs to decline further. So what factors have caused these reductions in soft costs?

An array of studies, particularly in the US have used highly detailed datasets to understand the sources of costs reductions in soft costs. First, installers have figured out ways to reduce the installed cost of a PV system. This installed cost includes the labor associated with mounting panels on a roof, electrically linking them, and integrating the PV systems with the household's electrical system and grid connection. Some of these improvements are associated with the notion of learning by doing (LbD), as workers perform repeated tasks, they find ways to get better at them (Bryan and Harter, 1899). Among a wide variety of empirical observations, this LbD effect has been observed in building aircraft (Benkard, 2000), ships (Thompson, 2001), and computer memory (Irwin and Klenow, 1994). It is not

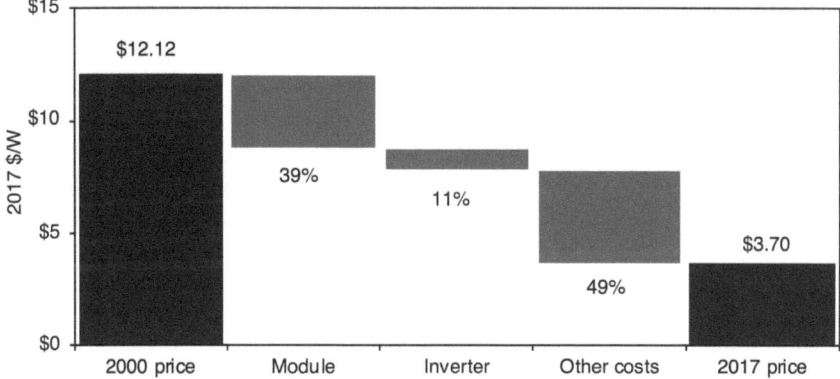

FIGURE 8.2 Sources of change in the prices of installed residential PV systems in the US.

difficult to picture workers on the roofs of houses finding it easier to mount a similar panel on a slightly different roof when doing it for the second time, or for the tenth time. This LbD effect extends to the range of activities involved with a system—not just the physical ones but also the administrative tasks of identifying potential customers, marketing to them, and securing a contract with them. While improvements from repetition provide one avenue for reducing costs, scale economies exist as well (Gillingham et al., 2016). As installers grow they attain scale that may allow them to save costs, for example by spreading fixed costs over a larger number of units. Scale may also allow them to invest in upgrading processes, such as training new employees rather than leaving it to them to develop their skills on the job site. A set of studies have found these learning from experience and scale effects repeatedly (Benthem et al., 2008; Nemet et al., 2017b). These and other studies have also found that there may be small-sized installers with low prices as well (O'Shaughnessy, 2018). These small-scale installers may need to set their prices low in order to enter the market and compete with installers that have more established reputations. Alternatively, they may have some economies of scope in that they are able to use complementary lines of business, such as roofing and electrical work, to acquire customers, manage billing, etc.

A second cost-reducing factor is that installers have stronger incentives to reduce their costs as markets become more competitive. Recent empirical work shows that as customers have gained access to better information about the available installer options, markets have become more competitive and prices have tended to fall (O'Shaughnessy and Margolis, 2018). Installers have even stronger incentives to reduce costs as they find themselves bidding against competitors. In the US, the rise of third-party quote providers like EnergySage have provided customers with better access to pricing information. O'Shaughnessy and Margolis (2018) test this hypothesis directly, showing that installers offer lower prices when they expect their customers to receive more quotes. It does seem however that the scale and learning effects are larger than the competition effects. As a result, markets with large established installers tend to have lower prices than ones with a large number of small- to medium-sized installers (Nemet et al., 2017b).

Third, there may also be some cost-reducing learning from experience on the demand side. As customers become more familiar with PV, due to its becoming more widespread, they may be more receptive to the idea of adopting it themselves (O'Shaughnessy, 2018). This effect may reduce the costs of acquiring customers for installers. A more direct channel is that potential adopters may find themselves with neighbors who have already gone through the process of installing PV and can provide information about attributes such as the reliability of installers and the pricing options available (Graziano and Gillingham, 2015). This effect may also reduce customer acquisition costs by making adopters less risk averse by being able to rely on the trusted information of people they already know (Nemet et al., 2017a).

Fourth, financing innovations have had a clear cost reducing effect on soft costs. The primary finance innovation in the US has been leasing arrangements, which are sometimes referred to as third-party ownership (TPO) systems, but which we will refer to here generally as leases. In leasing arrangements, homeowners lease a PV system from a company or enter into a power purchase agreement with a company for the electricity the PV system on their property produces (Davidson et al., 2015). In 2017, 43% of the 2013 residential installations were leased systems, while the remaining systems were customer owned (Barbose and Darghouth, 2018). These leased systems come in two basic varieties. In some cases, the third-party owner contracts with a separate entity to install the system, and the purchase price reported represents the payment from the third-party owner and the installation contractor. In other cases, however, the third-party owner conducts the installation itself, in which case no transaction occurs from which a purchase price can be identified. Leased systems are less expensive than customer owned systems and tend to have less variation in their pricing (Nemet et al., 2017b). The main impact, however, of leased systems is the reduction of up-front costs for capital constrained homeowners (Rai and Sigrin, 2013; Drury et al., 2014). A second financing innovation has been community shared solar, in which a large system is installed to provide power for hundreds of customers who subscribe to the installation (Chan et al., 2017). These systems provide economies of system scale in that they are in the order of hundreds of kilowatts instead of a few, thus saving on installation costs considerably (Funkhouser et al., 2015). Other financial innovations such as Property Assessed Clean Energy (PACE) financing, in which financing is provided by a municipality and paid through property taxes, have been found to increase adoption rates substantially, about double, when available (Kirkpatrick and Bennear, 2014). Other programs, known as "solarize," group purchases among a large set of homeowners to negotiate better equipment and installation prices.

Fifth, policies have affected soft costs as well. When installing a new system, installers must secure permits and then have the system inspected afterwards. These steps add costs and delays and vary across local jurisdictions. They affect costs directly and, in the jurisdictions with the most onerous procedures, can add hundreds of dollars to the costs of a residential system. Studies estimate that the most restrictive permitting adds $900 (Burkhardt et al., 2015), $1050 (Seel et al., 2014), and $2850 (Dong and Wiser, 2014) to the cost of an average sized five kW residential system. Consequently, efforts to streamline permitting, possibly enhanced by economies of scale in local program administration (e.g. hiring a specialist), can also reduce soft costs.

Another way in which policy can affect soft costs is through increasing demand for PV. A range of policies, such as rebates, FiTs, and tax credits enhance the demand for PV systems by reducing the effective price that adopters face. Subsidies shift the demand curve upward and increase the quantity and the price at which willingness to pay equals marginal cost. Thus, in the short-term, other things being equal, subsidies increase the prices of PV systems. However, subsidies also have a longer-term effect in that they provide opportunities for learning by doing and economies of scale. Since these are cost-reducing activities they enhance demand and obviate the need for continued subsidies. Thus, the overall effect of subsidies is to increase prices in the near term and reduce prices in the longer term.

A further consideration on the effect of demand side subsidies for PV is whether installers keep the subsidies for themselves—by passing on price increases to customers—or pass on the savings to their customers. The former would tend to dampen the learning curve effects of subsidy programs. While fortunately it is clear that installers are not able to retain the full value of subsidies themselves, the empirical evidence is split on how much of a subsidy households actually receive. One set of studies finds that households receive almost all the value of a subsidy (Dong et al., 2018; Pless and Benthem, 2017) while another set funds that installers retain a substantial share of the subsidies (Borenstein, 2017), that is, they raise prices when subsidies become available. As markets become more competitive and pricing more transparent, one expects customers to see an increasing number of subsidies.

Knowledge spillovers

While solar installers have learned from their experience in installing systems, they have also learned from the experience of other installers. Improved installation techniques and changes in business processes have diffused from one installer to another. When the know-how that one installer has gained from experiences becomes available to another, that knowledge is said to "spill over" between installers (Arrow, 1962; Ghemawat and Spence, 1985). Knowledge spillovers create both benefits and problems. They are favorable in that they disseminate knowledge across the industry. Rather than one firm developing a new process on its own and keeping that secret or patented, spillovers spread that innovation to other firms. While this process typically takes some time and the knowledge can also become obsolete over time, it distributes the beneficial effect of the improvement throughout the industry (Grubler and Nemet, 2014). Spillovers in PV soft costs have thus spread cost reductions throughout the industry which has helped stimulate adoption of the technology as prices have fallen.

The problem with knowledge spillovers is that they weaken incentives for investment in innovation. Once they are aware that knowledge spillovers are present in the industry, firms have an incentive to free ride on the experience of other firms, rather than developing saving innovations themselves. Assuming that these innovations involve some up-front costs—whether through R&D investment or forward pricing early installations to develop experience—firms will be more reluctant to make these investments if spillovers are present. Instead of the innovation giving that firm an advantage, competitors soon have access to that knowledge too, so the firm will be much less likely to invest in innovation and more likely to free ride on the learning investments of other firms. Taken to an extreme, all firms will delay deployment investments and wait for other firms to develop innovations that they can subsequently adopt at low cost.

We have seen knowledge spillovers in PV via an array of studies and multiple methodologies. Econometric studies of PV have sought to identify the effect of knowledge spillovers on the installed prices of PV systems by calculating the accumulated experience of each installer in the industry (Gillingham et al., 2016). These studies find that experience reduces costs and that the experience of other firms reduces costs, indicating that spillovers are real and significant (Nemet et al., 2017b). A geographic aspect is important as well since knowledge tends to move more rapidly over shorter distances. For example, one study of solar soft cost patent applications between 1997 and 2011 in the US found that local demand subsidies are significantly more effective in inducing local innovative activity than are national level policies (Rai and Beck, 2015). This inter-firm learning can occur through formal activities such as interactions at trade shows and professional education. It can also occur more informally thorough the assimilation of new employees from other firms, observing the activities of other firms, speaking with the customers of competitors, and social networking with employees of other firms.

Knowledge spillovers create what is known as a "positive externality" and thus constitute a market failure. That is, prices, supply, and demand will produce a suboptimal amount of PV deployment—because firms will be waiting around to free ride and will be less innovative than the optimal case (Nemet, 2012). The policy remedy to correct this market failure is to subsidize demand for PV so that the optimal level of deployment and innovation can occur. Setting the levels of subsidies has proven to be a challenging task, as examples in California (Benthem et al., 2008) and Germany (Unnerstal, 2017) show. But the evidence showing that knowledge spillovers are present in the PV industry provides a justification for subsidizing PV, even if the avoided pollution from PV is accounted in some other form, for example via a tax on pollution.

Emerging challenges and opportunities

Now that PV has grown beyond niche markets, it has encountered confrontations with issues that in the recent past would have been dismissed as good problems to have. Three emerging challenges are of particular importance: intermittence, utility business models, and financing in developing countries. All of them have local roots and will likely require heterogeneous efforts across the globe to resolve. First, it is not sunny all the time. The industry jargon—especially among those opposed to solar—for the effects of clouds, night, winter, and occasional eclipses, is that solar is "intermittent" (Hirth, 2015). Solar advocates use the term "variable." In any case, unlike fossil fuels and nuclear, solar electricity may not be available when people need electricity. Whereas excessive costs used to be the primary reason that serious energy analysts could not take solar seriously, the dissolution of that argument has focused attention on intermittence. Solar is risky, even dangerous. Relying on solar would risk blackouts. A more nuanced view is that the value of solar will decline as more and more of it is used, because of the increased need to invest in back up generation and deploy other means of accommodating its use (Sivaram, 2018). Solar's intermittence is a central argument from those who oppose solar in addition to the legitimate concern about the costs of accommodating solar once it reaches large levels of energy supply. However, we have seen this same emphasis on innovation and flexibility to overcome bottlenecks in making solar panels, as well as other areas. That provides grounds to be optimistic that intermittence is just another problem to be solved, not a deal-breaker by any means. This optimism derives from existence of a plethora of possible solutions: battery storage, flywheel storage, compressed air storage, thermal storage, electric vehicles as storage, flexible demand response, and using electricity for non-time-dependent activities like electrifying industrial processes (like steel), desalination, hydrogen production, and capture of carbon dioxide from air. A narrow view assuming that PV needs to be adapted to behave as much like a fossil fuel power plant as possible misses the point and misses these opportunities.

Second, the threat of a utility death spiral is real. On September 13, 2012 at its board meeting in Colorado Springs the Edison Electric Institute, the industry group for the US's electric utilities, began sounding the alarm about solar because of

> the opportunity for powerful, agile, and unregulated companies to get into the business that utilities have traditionally seen as near-exclusively their own. (Craver et al., 2012)

Contracts were then being signed at below $100 per MWh. A few months later the group published a report arguing that the way people pay for electricity needed to change because of this threat (Ackerman and Martini, 2013). To make the stakes clear, the report used the cautionary tale of the landline telephone industry in the 1990s, which entered a "death spiral," in which it raised prices to make up for a dwindling customer base, which shrank further as it encountered higher prices. Indeed, the landline industry saw its revenues halve even as the size of the telecom industry doubled due to mobile phones, e-commerce, and online content. Watching the four large coal-based electric utilities in Germany lose billions and risk bankruptcy as renewables drove prices lower provided further evidence that US electric utilities were about to be "disrupted."

A few strategies to avoid the fate of landlines and German utilities have emerged. One prevalent strategy has been to lobby regulators for a different way of charging customers, specifically, to charge a large fixed amount per month rather than per unit of electricity. This strategy was first tried in Wisconsin in the US and was ultimately successful. But in the end, such tactics only delay the death spiral, as it encourages customers to move even further from the utility system. More proactive models are emerging, even as the threat of disruption continues to loom (Mills et al., 2016). As a result, the speed at which PV becomes widely adopted will depend on the political economy context in which it operates. Large entrenched interests have a lot to say about PV's future. Whether they succeed in blocking PV or embrace it and adapt their business models to it will play a large role in the speed at which PV is adopted. The latter will require substantial changes by electric utilities as well as the regulators charged with determining how utilities are compensated for their costs and investments in proving power.

Third, the need for financing is greatest where people need power most. While most solar deployment to date has been in developed countries, almost all of the growth in energy demand in the next two decades will come from developing countries: India, China, and countries in Africa, the Middle East, and Southeast Asia. For PV to succeed in those places it will need to be cheap, thus the evolution described here is central. However, financing new PV can be challenging. PV is what is known as "capital intensive." That does not mean that PV is expensive. Rather, it refers to the timing of costs; it means that most of the costs of PV occur at the beginning of its lifetime. That makes it distinct from coal and gas power plants where, to a large extent, you pay as you go, in the form of on-going payments for fuel and workers. While not having to pay for fuel or maintenance is a tremendous asset of solar, it does require adopters to come up with an up-front payment. That can be challenging in countries with capital markets that are not as well developed, i.e. where it is more difficult to borrow money. For example, whereas a homeowner in a developed country can borrow at 3–4%, rates in developing countries can exceed 10% (Ondraczek et al., 2015). Just this change in financing would double the lifetime cost of solar. Successful examples of overcoming these challenges include: more stable regulatory environments, innovative financing programs, and capacity building to develop local expertise (Creutzig et al., 2017).

While the widespread adoption of PV is raising new challenges, it is also creating new opportunities. The focus of the frontier of PV know-how has shifted over time—from the science to devices to manufacturing to the equipment used for manufacturing. As PV becomes cheaper still, the knowledge frontier appears to be shifting to how people use PV. We have seen this already in the sequence of niche markets that began with satellites in 1957. But as PV becomes very inexpensive new creative applications are emerging. One example is that manufacturing is starting to become less concentrated. Even as China accounts for over two-thirds of module production, emerging users of PV panels such as India, Egypt, and Turkey, are beginning to locate parts of the PV value chain to their countries to serve domestic markets. Turnkey and turnkey-light process lines are enabling rapid scale up of production that can be cost effective even without the ten-gigawatt scale that the giants in China are approaching. Another example is the increasing integration of storage with PV, and in some cases integrated with electric vehicles. The falling cost of batteries is enabling cost-effective storage over multiple hours, with half of new systems in Germany now including batteries. The advent of vehicle-to-grid service companies, which provide voltage frequency regulation from cars for the electric grid further enables this integration. Combining PV with direct air capture (Chapter 10) provides another new application, in that case to remove CO_2 from the atmosphere. Countries like Saudi Arabia are considering massive investment that would take even further advantage of their favorable combination of rapidly growing electricity demand, low cost financing, plentiful non-arable land, and abundant solar radiation. While one application would be to provide power that would divert oil to exports rather than for domestic electricity production another would be to use PV to power new city centers, such as Neom. More ambitiously, the development of hundreds of gigawatts of very low-priced solar electricity to the Middle East could shift energy intensive manufacturing activities to sunny regions, where electrically produced process heat and power itself would be abundant.

Conclusion

Soft costs have fallen substantially over the past two decades, even if not as dramatically as the prices of modules. Soft costs have fallen due to an array of actions, almost none of which involve manufacturing know-how and automation, the key drivers of module price declines. Rather these declines involve intrinsically human elements, such as the acquisition of tacit knowledge in the heads of those wielding wrenches on one roof after another. Above all, soft costs have fallen due to this broader process of learning by doing, the improvement of installation that has occurred as installers have gained experience. An important characteristic of this learning is that it has spilled over—employees have learned from each other, and from other firms. Those spillovers are important justifications for PV subsidies in that without subsidies there would be too few installations and firms would free ride on the experience paid for by other firms. Like other cases of learning by doing spillovers, that process diminishes with distance. That makes learning an inherently local process, which emphasizes the importance local policies in addition to national ones.

Locally constrained learning has also engendered political support for solar—even when the production of modules was clearly moving far outside those jurisdictions, mostly to China. As soft costs have become larger than module costs, the local value being added to solar has effectively outweighed the forces of globalization that have moved production value far away. We have yet to see any substantial multi-national residential PV installers, even if some globally present corporations have done system integration for utility scale installations. As discussed above, this inherently local attribute of PV installers has created a stable source of political support. The combination of a growing cohort of PV adopters and of installer firms has created an advocacy coalition that has been crucial to the development of local PV support, some of which has led to national support. Early adopters and installers were central to the bottom-up PV movement in Germany, which began in Aachen in 1992 and took off on the federal level in 2000. Even though Germany has lost much of its PV manufacturing employment to China, it still retains its roofers, electricians, and solar installation specialists. Something similar occurred in California as well, in a small way in the 1980s and in a big way after 2006, with federal policy to follow. We do not see this same bottom-up movement in Japan and China despite some of the same locally oriented adoption and learning dynamics there. Although even in Japan and China some small early success in PV installs was used to justify larger programs later on.

References

Ackerman, E. & Martini, P. D. 2013. *Future of Retail Rate Design*. Edison Electric Institute.

Ardani, K. & Margolis, R. 2015. *Decreasing Soft Costs for Solar Photovoltaics by Improving the Interconnection Process: A Case Study of Pacific Gas and Electric*. Golden, CO, National Renewable Energy Laboratory (NREL).

Arrow, K. 1962. The economic implications of learning by doing. *The Review of Economic Studies*, 29, 155–173.

Barbose, G. L. & Darghouth, N. R. 2018. *Tracking the Sun: Installed Price Trends for Distriuted Photovoltaic Systems in the United States*. Berkeley, CA, Lawrence Berkeley National Laboratory.

Barbose, G. L., Darghouth, N. R., Millstein, D., Lacommare, K. H., Disanti, N. & Widiss, R. 2017. *Tracking the Sun 10: The Installed Price of Residential and Non-Residential Photovoltaic Systems in the United States*, Lawrence Berkeley National Laboratory and SunShot

Benkard, C. L. 2000. Learning and forgetting: The dynamics of aircraft production. *American Economic Review*, 90, 1034–1054.

Benthem, A. V., Gillingham, K. & Sweeney, J. 2008. Learning-by-doing and the optimal solar policy in California. *The Energy Journal*, 29, 131–152.

Bollinger, B. & Gillingham, K. 2012. Peer effects in the diffusion of solar photovoltaic panels. *Marketing Science*, 31, 900–912.

Borenstein, S. 2017. Private net benefits of residential solar PV: The role of electricity tariffs, tax incentives, and rebates. *Journal of the Association of Environmental and Resource Economists*, 4, S85–S122.

Bryan, W. L. & Harter, N. 1899. Studies on the telegraphic language: The acquisition of a hierarchy of habits. *Psychological Review*, 6, 345–375.

Burkhardt, J., Wiser, R., Darghouth, N., Dong, C. G. & Huneycutt, J. 2015. Exploring the impact of permitting and local regulatory processes on residential solar prices in the United States. *Energy Policy*, 78, 102–112.

Chan, G., Evans, I., Grimley, M., Ihde, B. & Mazumder, P. 2017. Design choices and equity implications of community shared solar. *The Electricity Journal*, 30, 37–41.

Craver, T. F., Piro, J. J & Wolff, B. 2012. Electric Transportation. Edison Electric Institute Fall Board and Chief Executives Meeting, September 2012 Colorado Springs.

Creutzig, F., Agoston, P., Goldschmidt, J. C., Luderer, G., Nemet, G. & Pietzcker, R. C. 2017. The underestimated potential of solar energy to mitigate climate change. *Nature Energy*, 2, nenergy2017140.

Davidson, C., Steinberg, D. & Margolis, R. 2015. Exploring the market for third-party-owned residential photovoltaic systems: Insights from lease and power-purchase agreement contract structures and costs in California. *Environmental Research Letters*, 10, 024006.

Dong, C. & Wiser, R. 2014. The impact of city-level permitting processes on residential photovoltaic installation prices and development times: An empirical analysis of solar systems in California cities. *Energy Policy*, 63, 531–542.

Dong, C., Wiser, R. & Rai, V. 2018. Incentive pass-through for residential solar systems in California. *Energy Economics*, 72, 154–165.

Drury, E., Jenkin, T., Jordan, D. & Margolis, R. 2014. Photovoltaic investment risk and uncertainty for residential customers. *IEEE Journal of Photovoltaics*, 4, 278–284.

Friedman, B., Ardani, K., Feldman, D., Citron, R., Margolis, R. & Zuboy, J. 2013. *Benchmarking Non-Hardware Balance-of-System (Soft) Costs for US Photovoltaic Systems Using a Bottom-Up Approach and Installer Survey*. NREL.

Fu, R., Feldman, D. J., Margolis, R. M., Woodhouse, M. A. & Ardani, K. B. 2017. *US solar photovoltaic system cost benchmark: Q1 2017*. Golden, CO, National Renewable Energy Lab. (NREL).

Funkhouser, E., Blackburn, G., Magee, C. & Rai, V. 2015. Business model innovations for deploying distributed generation: The emerging landscape of community solar in the US. *Energy Research & Social Science*, 10, 90–101.

Ghemawat, P. & Spence, A. M. 1985. Learning-curve spillovers and market performance. *Quarterly Journal of Economics*, 100, 839–852.

Gillingham, K., Deng, H., Wiser, R. H., Darghouth, N., Nemet, G., Barbose, G. L., Rai, V. & Dong, C. 2016. Deconstructing solar photovoltaic pricing: The role of market structure, technology, and policy. *The Energy Journal*, 37, 231–250.

Graziano, M. & Gillingham, K. 2015. Spatial patterns of solar photovoltaic system adoption: The influence of neighbors and the built environment. *Journal of Economic Geography*, 15, 815–839.

Grubler, A. & Nemet, G. F. 2014. Sources and consequences of knowledge depreciation. In: Grubler, A. & Wilson, C. (eds.) *Energy Technology Innovation: Learning from Historical Successes and Failures*. Cambridge, Cambridge University Press.

Hirth, L. 2015. The optimal share of variable renewables: How the variability of wind and solar power affects their welfare-optimal deployment. *The Energy Journal*, 36.

Irwin, D. A. & Klenow, P. J. 1994. Learning-by-doing spillovers in the semiconductor industry. *Journal of Political Economy*, 102, 1200–1227.

Karakaya, E., Nuur, C. & Hidalgo, A. 2016. Business model challenge: Lessons from a local solar company. *Renewable Energy*, 85, 1026–1035.

Kirkpatrick, A. J. & Bennear, L. S. 2014. Promoting clean energy investment: An empirical analysis of property assessed clean energy. *Journal of Environmental Economics and Management*, 68, 357–375.

Krause, R. M., Carley, S. R., Warren, D. C., Rupp, J. A. & Graham, J. D. 2014. "Not in (or under) my backyard": Geographic proximity and public acceptance of carbon capture and storage facilities. *Risk Analysis*, 3, 529–540.

Lee, C.-Y. & Heo, H. 2016. Estimating willingness to pay for renewable energy in South Korea using the contingent valuation method. *Energy Policy*, 94, 150–156.

Mills, A., Barbose, G., Seel, J., Dong, C., Mai, T., Sigrin, B. & Zuboy, J. 2016. *Planning for a Distributed Disruption: Innovative Practices for Incorporating*, Berkeley, CA, Ernest Orlando Lawrence Berkeley National Laboratory.

Moezzi, M., Ingle, A., Lutzenhiser, L. & Sigrin, B. 2017. *A Non-Modeling Exploration of Residential Solar Photovoltaic (PV) Adoption and Non-Adoption. National Renewable Energy Laboratory (NREL)*, Golden, CO, National Renewable Energy Laboratory. NREL/SR-6A20–67727.

Nemet, G. F. 2012. Subsidies for new technologies and knowledge spillovers from learning by doing. *Journal of Policy Analysis and Management*, 31, 601–622.

Nemet, G., O'Shaughnessy, E., Wiser, R. H., Darghouth, N. R., Barbose, G. L., Kenneth, G. & Rai, V. 2016. *Characteristics of Low-Priced Solar Photovoltaic Systems in the United States*, Berkeley, CA, Lawrence Berkeley National Laboratory.

Nemet, G. F., O'Shaughnessy, E., Wiser, R., Darghouth, N. R., Barbose, G., Gillingham, K. & Rai, V. 2017a. What factors affect the prices of low-priced U.S. solar PV systems? *Renewable Energy*, 114, 1333–1339.

Nemet, G. F., O'Shaughnessy, E., Wiser, R., Darghouth, N., Barbose, G., Gillingham, K. & Rai, V. 2017b. Characteristics of low-priced solar PV systems in the U.S. *Applied Energy*, 187, 501–513.

Nomura, N. & Akai, M. 2004. Willingness to pay for green electricity in Japan as estimated through contingent valuation method. *Applied Energy*, 78, 453–463.

Nyborg, K., Anderies, J. M., Dannenberg, A., Lindahl, T., Schill, C., Schlüter, M., Adger, W. N., Arrow, K. J., Barrett, S., Carpenter, S., Chapin, F. S., Crépin, A.-S., Daily, G.,

Ehrlich, P., Folke, C., Jager, W., Kautsky, N., Levin, S. A., Madsen, O. J., Polasky, S., Scheffer, M., Walker, B., Weber, E. U., Wilen, J., Xepapadeas, A. & De Zeeuw, A. 2016. Social norms as solutions. *Science*, 354, 42–43.

O'Shaughnessy, E. 2018. *Solar Photovoltaic Market Structure in the United States: The Installation Industry, Effects on Prices, and the Role of Public Policy*. PhD thesis, University of Wisconsin-Madison.

O'Shaughnessy, E. & Margolis, R. 2018. The value of price transparency in residential solar photovoltaic markets. *Energy Policy*, 117, 406–412.

Ondraczek, J., Komendantova, N. & Patt, A. 2015. WACC the dog: The effect of financing costs on the levelized cost of solar PV power. Renewable Energy, 75, 888–898.

Pless, J. & Benthem, A. A. V. 2017. *The Surprising Pass-Through of Solar Subsidies*. National Bureau of Economic Research Working Paper Series, No. 23260.

Rai, V. & Beck, A. L. 2015. Public perceptions and information gaps in solar energy in Texas. *Environmental Research Letters*, 10, 074011.

Rai, V. & Robinson, S. A. 2013. Effective information channels for reducing costs of environmentally-friendly technologies: Evidence from residential PV markets. *Environmental Research Letters*, 8, 014044.

Rai, V. & Sigrin, B. 2013. Diffusion of environmentally-friendly energy technologies: Buy versus lease differences in residential PV markets. *Environmental Research Letters*, 8, 014022.

Rand, J. & Hoen, B. 2017. Thirty years of North American wind energy acceptance research: What have we learned? *Energy Research & Social Science*, 29, 135–148.

Seel, J., Barbose, G. L. & Wiser, R. H. 2014. An analysis of residential PV system price differences between the United States and Germany. *Energy Policy*, 69, 216–226.

Sivaram, V. 2018. *Taming the Sun: Innovations to Harness Solar Energy and Power the Planet*, Cambridge, MA, The MIT Press.

Sommerfeld, J., Buys, L. & Vine, D. 2017. Residential consumers' experiences in the adoption and use of solar PV. *Energy Policy*, 105, 10–16.

Sundt, S. & Rehdanz, K. 2015. Consumers' willingness to pay for green electricity: A meta-analysis of the literature. *Energy Economics*, 51, 1–8.

Thompson, P. 2001. How much did the Liberty shipbuilders learn? New evidence for an old case study. *Journal of Political Economy*, 109, 103–137.

Unnerstal, T. 2017. *The German Energy Transition: Design, Implementation, Costs, and Lessons*, Springer.

Urpelainen, J. & Yoon, S. 2015. Solar home systems for rural India: Survey evidence on awareness and willingness to pay from Uttar Pradesh. *Energy for Sustainable Development*, 24, 70–78.

PART 4
Doing it again

9

SOLAR AS A MODEL TO FOLLOW

The evolution of solar technology from a scientific curiosity to a scalable source of low-carbon energy supply has provided society with a powerful tool with which to affordably contribute to a decarbonized world economy. But the challenge of addressing climate change requires such a large transformation of the energy system that even massive deployment of PV will be insufficient. Part of the value in solar's successful evolution is that it provides a model for the development of other low-carbon technologies that society will need to meaningfully address climate change. This last section of the book describes how solar can serve as a model. Understanding solar's progress is the key to making smart decisions about developing climate technologies we will need. The utility of solar as a model relies on answering three questions: 1) What were the drivers of solar's success? 2) What attributes of PV technology made it amenable to those drivers? And 3) To what types of technologies might this model be applied? The characteristics of slow change in the climate system and in the energy system mean that the solar model will only be useful if it can be accelerated. This chapter concludes by proposing nine approaches to accelerating low-carbon innovation for technologies that are close analogues to PV.

Why was solar successful?

The preceding eight chapters describe the evolution of PV technology over 180 years from Becquerel's discovery in 1839 to the low cost, massive scale PV production we see today. Geographically and temporally, PV's evolution can be summarized as the result of distinct contributions by the US, Japan, Germany, Australia, and China—in that sequence. Summarizing this development pathway, PV improved as the result of:

1. Scientific contributions in the 1800s and early 1900s, in Europe and the US, that provided a fundamental understanding of the ways that light interacts with molecular structures (Chapter 3);
2. A breakthrough at a corporate laboratory in the US in 1954 that made a commercially available PV device (Chapter 3);
3. A major government R&D and public procurement effort in the 1970s in the US (Chapter 4);
4. Japanese electronic conglomerates serving niche markets in the 1980s and launching the world's first major rooftop subsidy program in 1994 with a declining rebate schedule (Chapter 5);
5. Germany adopting a feed-in tariff in 2000 that quadrupled the market for PV and developing production equipment that automated and scaled PV manufacturing (Chapter 6);
6. Entrepreneurs, many trained in Australia, building factories in China of gigawatt scale in the 2000s and creating the world's largest market for PV from 2013 onward (Chapter 7); and
7. A cohort of adopters with high willingness to pay, accessing information from neighbors, and installer firms that learned from their installation experience, as well as that of their competitors to lower soft costs (Chapter 8).

These factors led the price of PV to fall by two orders of magnitude, grow at over 30% per year for close to three decades, and resulted in a long-term learning rate above 20%.

As we search for generalizable aspects of PV's success, it helps to think of these specific historical developments in the context of an innovation framework (Gallagher et al., 2012) that can help attain some external validity. That is, if we can connect these events to theory, we are in a much better position to apply them to other technologies—which is the ultimate goal of this project. Beyond these historical characteristics are nine broad drivers of PV's cost reductions:

Scientific understanding of a phenomenon: The advent of the first efficient solar cell in 1954 by Bell Labs depended on understanding and producing the p–n junction, which enabled the creation of a voltage potential on two sides of a semi-conductor device. That understanding derived from Einstein's theory of the photoelectric effect and how light interacted with the atomic structure. Importantly, the theory was fully transparent as scientific knowledge and freely available to those who could access the 1905 paper and understand its equations. The knowledge could thus disseminate widely and quickly.

Evolving R&D foci: PV R&D traces back to work on early silicon photoelectric devices in the 1930s at Siemens and continues today. R&D has been effective for PV at various times and locations—in the 1930s and 1940s when the discovery of the p–n junction opened a line of investigation; in the 1950s as the space race heated up after Sputnik; in the 1970s in the US and Japan when governments responded to the oil crises of 1973 and 1979 with infusions of public R&D that generated a burst of breakthroughs; from the mid-1980s onward at the University of New South Wales; in the 2000s in Germany when equipment manufacturers responded to the enlarged market by developing dedicated PV machinery; and more recently as investigations into perovskites, quantum dots, and hybrids with silicon crystals in a variety of locations. For the most part, the knowledge generated by these investments, both public and private, flowed globally and quickly. The machine manufacturers were perhaps the strongest at retaining the full value of their R&D investments, but that knowledge still dispersed because the machines were mobile and the firms producing them successfully marketed them around the world.

Iterative upscaling: From its origins, the massive resource on which PV depends, sunlight hitting the Earth, has engendered ideas about the large role PV could play in the energy system. Growth and scale have been central concerns since the beginning of practical solar in the 1950s. As commercialization was taken more seriously in the 1970s, the benefits of manufacturing scale became widely apparent. Expectations were widespread that larger manufacturing facilities were a central means by which to reduce costs. But even though people in the industry have known for decades that cost would fall as production scale grew, increases in scale occurred only gradually. Production by the biggest producer in the world has increased by a factor of 1,000 between 1988 and 2018, growing at about 30% per year. The iterative aspect of this gradual scaling process gave the manufacturers time to make improvements along the way—either to adjust to new difficulties that emerged at larger scale or to take advantage of new opportunities that emerged at scale, such as automation.

Learning by doing: This latter aspect of upscaling involved substantial learning-by-doing. As manufacturers gained experience in PV production, they found ways to improve the process and reduce costs. This thinking goes back to the US Block Buy in the 1970s, and informed various programs thereafter including the Japanese rooftop program in the 1990s, the German FiT in the 2000s, the California Solar Initiative in 2006, and Chinese entrepreneurs in the 2000s who knew that they could reduce costs as they gained experience in production. Moreover, the other parts of the PV value chain, such as installing panels on roofs, also improved from experience.

Knowledge spillovers: Crucially, the spillovers from learned experience provided the opportunity for other firms and countries to observe, take advantage of, and build upon these lessons. These so-called knowledge spillovers have contributed to cost reductions and have helped justify the subsidy programs that have been so important to PV's growth.

Modular scale: Like other energy technologies, economies of scale have been central to PV. However, in contrast to many other energy technologies, these economies of scale have been most influential in manufacturing rather than producing larger units. The initial cells on the Vanguard satellite and those used to power consumer electronics in the 1980s were postage stamp scale, with less than a watt of capacity. Contrast these with coal or nuclear power plants for which hundreds of millions of watts are a minimum scale. Household panels were bigger, but still only a couple of 100 Watts. Panels could be strung together to make systems of any size. Since the economies of scale were strong in manufacturing, it was affordable to produce cells at 100-Watt scale. That size meant that they could be used in the variety of niche markets that were so important to solar. More recently economies of scale have been found in unit scale (the size of individual installations) as utility scale projects show lower costs than rooftop ones.

Policy-independent niche markets: The modular scale of PV allowed it to serve a sequence of increasingly large niche markets with decreasing willingness to pay: satellites, marine buoys, lighting on oil rigs, telecom repeater stations, green off-grid consumers, and later utility scale installations of 100s of MWs. Each niche market allowed producers to scale up production and learn by doing. The technology became demonstrated, familiar, and reliable. Most of these markets were policy independent, so when PV was out of favor with policymakers—in the 1950s with the rise of nuclear and in the 1980s with low energy prices—these niche markets kept the industry alive, retained the tacit knowledge in people's heads, and allowed the industry to grow until the next policy window opened.

Robust policy support: When supportive policies abruptly ended, PV survived not just by serving policy-independent niche markets, but also by serving new jurisdictions. When the Japanese rooftop program ended in 2005, the German market was exploding; when the Spanish market collapsed in 2009, the German market remained strong; when the German market shrank after 2012, the Chinese FiT created the biggest national PV market ever. In other cases, sub-national policies, such as in German cities and in California, bolstered markets when national policies were absent. These markets overlapped in time and space and provided robustness to the expectations of future markets, that is, if one policy abruptly cancelled, other drivers would sustain markets. Moreover, the increasingly local aspect of so much of the PV value chain has strengthened political support, even as globalization has dispersed the manufacturing activities. Expectations have been so crucial to PV's success that having robust markets was enormously important for creating strong incentives to invest in the technology for the future, whether through R&D, training employees, forward pricing, or capacity expansion.

Delayed system integration challenges: As discussed in Chapter 8, PV is confronting an array of challenges as it moves to larger shares of electricity demand, such as integration into electric grids and the modernization of utility business models and regulations. While these are serious challenges, they are only becoming ones that might limit PV's growth in the late 2010s. For the previous 70 years, PV was able to grow, improve, and get cheaper without concern about the availability of storage and reluctance of electric utilities. PV can now address these challenges because of the tremendous benefits from complementary technologies such as inexpensive storage, a smarter grid, and novel financing schemes. It also does this from a much stronger position, aided by well-established advocacy coalitions that can mobilize political support. The ability to delay the confrontation of system integration challenges has positioned PV to more effectively address them in the future. Contrast this timing with that of large-scale technologies like nuclear power, and carbon capture and storage in which system integration challenges, in those cases at the plant level, are encountered with the production of the first unit. In the case of both of those technologies, system integration has led to enormous costs and threatens to choke off growth of the industry. At the same time, these industries are learning how to price at scale, trying to accumulate learning by doing, and expanding production scale. PV has been in a much stronger position to confront system integration by being able to wait until the industry is mature to do so.

Characteristics of PV

Other technologies have been in a position to grow quickly, to improve, and to fall in costs, but many, if not most, have disappointed. What is it about solar that made it amenable to these drivers described above? I suggest that the following characteristics of PV made it amenable to the drivers of cost-reductions above. They fit in three categories: characteristics of the device itself, of the way it is produced, and of the people adopting it.

Device

Key characteristics of the PV device itself have facilitated the path it has taken. First, PV exploits a *scientific phenomenon*. That characteristic is one that could lead one to call it high-technology. The strong link to science helped make R&D investments pay off. It made discoveries into budding blocks because they had general applicability in that they fit into a framework of understanding, for example how p–n junctions work and how light interacts with solid matter.

Second, the PV industry converged on a *dominant design* around 1985. It spent the next thirty years optimizing that design and it turned out there was plenty of room for improvements. At the same time, alternative designs emerged—such as thin film that challenged crystallized silicon (x-Si) PV and raised the bar (i.e. lowered the cost threshold) so that x-Si kept improving. That design also provided opportunities for further improvement. A long-term asset for PV is the possibility of bringing in new materials which can further extend cost reductions. While some fret that x-Si may be a dead end and is preventing novel PV technologies from competing a (Sivaram et al., 2018), a dominant design is typically essential for growth. Hybrid designs, e.g. perovskites on x-Si, can provide a transition path to take advantage of the massive scale of x-Si with the appealing attributes of low cost, high efficiency new materials that can boost its performance.

Third, a consequence of this convergence is that PV became a *standardized product*. Units could be configured in many ways. One does not need specialized machinery to put PV on a roof and it is relatively easy to integrate PV into the existing technological system. We do not need a new infrastructure to accommodate large amounts of PV. Roofers and electricians could enter the installation businesses quickly and easily. A grid connection can be made by a licensed electrician. Individuals could even do it themselves, and about 1% of installs are done that way in the US.

Fourth, PV technology, production, and companies had an openness to *technology spillovers*. PV could benefit from developments in related areas, particularly the computer industry but also more far afield sectors, such as radial tires that produced the wires needed for sawing wafers. Further, PV made use of the grid integration and policy design progress that wind power had initiated when wind was more affordable.

Production process

The way that PV has been produced also provides a set of distinguishing characteristics. First, more than any other energy supply technology, manufacturing PV has involved *massive units of production*. Assuming a 150 W panel, about four billion panels have been manufactured in the course of the industry and on the order of a 100 billion solar cells have been produced. This has provided a wide array of opportunities for innovation, most of which relate to the number of opportunities for iterative improvement that such volume provides.

Second, PV production has proven highly suitable for *automation*. As a result, the technology frontier shifted from product innovation to process innovation. The innovators in the industry shifted from cell designs to equipment production. That equipment became widely available as innovation was not retained by a single producer, but rather dispersed among many. This likely enabled a more competitive and less concentrated manufacturing industry, as shown in Chapter 7.

Third, tolerance for design compromises enabled PV to pursue *disruptive innovation*. A PV panel is not mission critical to its users; if one panel fails that likely only affects a small portion of the output and it can easily be replaced. Further, other than a few niche applications, one need not optimize electrical conversion efficiency. The initial round of major cost-reducing breakthroughs occurred in the late 1970s with the first cells designed for terrestrial applications rather than the extremely narrow tolerances, reliability, and efficiency required for space. Later, the Chinese in the 2000s took this disruptive approach to manufacturing even further by making the right compromises on manufacturing, such that a small sacrifice in efficiency, on the order of 1–2%, could enable very large reductions in cost.

Fourth, the investment to enter the market in producing PV is not prohibitive. The investment to get started is relatively low and the know-how required modest now that manufacturing production is so well established, and to a large extent turnkey. As a result, despite there being massive economies of scale in manufacturing, production is not very concentrated and thus extremely competitive—even more so than it was 15 years ago. This ability to enter has allowed entrepreneurs to play a role in the US from the 1950s–1980s, later in Germany, and especially in China in the 2000s.

Adoption

Characteristics related to how PV has been adopted also define PV's development path. First, PV's modular design, which allows it to be affordable at small scale, has many benefits. Foremost it enables PV to serve an array of niche applications—from calculators to rooftops to massive utility scale installations. It may have been good fortune that these niches existed at all, but is also possible that the flexible scale of PV made people realize that there was actually a market for power in these applications. PV's easily adaptable scale may have created these opportunities.

Second, PV appeals to large portions of the public. In most cases it need not fight a battle for public acceptance. In contrast, people use solar panels on their house to publicly display aspects of identity. Neighbors learned from each other and were able to depend on trusted sources when installing PV. The outcome was clusters of adopters, similar to an epidemiological disease model of technology diffusion. A consequence is strong local value-added that helps maintain political support for PV.

Third, many aspects of PV are geographically mobile. Because wafers, cells, and panels are small, they can be shipped easily, which is important because there are many locations suitable for solar. Even less sunny places like Germany, with a third the solar resource of the sunniest places on Earth, have been important. This footloose aspect has made the technology well-suited to international markets, trade, and geographically dispersed supply chains. It has also made the PV industry nimble to adjust to changing policies in various jurisdictions, as well as changing comparative advantage.

Other technologies share these characteristics of the device, the production process, and adoption. They thus seem amenable to taking advantage of one or more of the drivers that have been important to cost reductions in solar. This configuration of characteristics has been important for making solar amenable to the drivers that have led to massive improvement over decades.

Solar was too slow

Even though solar has been an immense success, it still provides insufficient guidance for addressing climate change because it took far too long to achieve widespread adoption. For example, one could use the Bell Labs Solar Battery in 1954 as a starting point of solar's path because it represented the first commercial demonstration of a solar cell. Widespread adoption of PV would generally require at least 10% of global electricity supply, which remains for the future. However, one can look at milestones that show it is well on its way to heading to widespread levels of adoption. Solar first reached 1% of electricity generation in a major country in 2009 in Germany (Figure 1.2). Solar first became low cost when bids came in below $20 per MWh in 2017. It thus took between 55 and 63 years for solar to go from a viable commercial technology to inexpensive and entering the realm of widespread adoption. We do not have that much time on climate change.

Modeling of the 2.0°C and 1.5°C temperature targets central to the Paris Agreement on climate change (UNFCCC, 2015) implies rapid decarbonization of the world almost immediately (Fuss et al., 2018) and massive deployment of negative emissions technologies in the 2030s (Nemet et al., 2018).

Unfortunately, solar is not the only technology that has taken a long time to move from laboratory to widespread adoption. The notion of formative phases comes from the innovation systems literature that analyzes the conditions required for technology development and adoption to be successful (Hekkert et al., 2007). Such literature emphasizes the health of the functions in an innovation system, such as mobilization of resources (Bergek et al., 2008). Motivated by the urgency of technologies needed for climate change, more recent work has sought to make the notion of formative phases more operational so it can be measured (Bento et al., 2018). A technology's formative phase begins with its "first embodiment," i.e. the first prototype of demonstration (Bento and Wilson, 2016). For PV, the Bell announcement in 1954 corresponds perfectly. The formative phases ends when 10% of the eventual saturation level is reached (Bento and Wilson, 2016). One could say that PV saturation is at 50% of electricity supply (Creutzig et al., 2017), so a 5% market share would be the end of the formative phases, corresponding to 2014 in Germany, and not yet attained for the world. Thus, PV's formative phase has taken sixty years.

Research on an array of other technologies has looked at the length of "formative phases" in technology development and found a mean of 22 years with a range of 4 to 85 years across technologies (Bento and Wilson, 2016). The upper range is twice as high if using stricter definitions of beginning and end times. Moreover, the post-formative phase period also takes time. Diffusion times, the time required for a technology to proceed from 10% of its eventual saturation level to 90% of that level, also takes decades, 20–60 years (Wilson, 2012; Grubler et al., 2016). So even though PV's evolution has been striking, it has been slower than average.

As an example, consider direct air carbon capture and sequestration (DAC). In DAC, ambient air is reacted with chemicals to absorb CO_2, which is then heated and reacted further to separate it out and store. The first commercial installation was to supply CO_2 to a greenhouse in Switzerland and came on-line in 2017. The first commercial PV installation was on the Vanguard Test Vehicle 3, whose rocket engines failed at launch in December 1957. If one were to apply the same 60-year time period it took from the first commercial PV installation, the Vanguard test satellite in 1957 to inexpensive solar in 2017, we can expect low cost DAC in 2077. If we are to have widespread deployment of PV in 2040, then we can expect to wait another widespread carbon removal via DAC by 2100 (Figure 9.1). That is far too late to make much difference in avoiding the damages from climate change (Nemet et al., 2018).

The long time frames included here suggest that new technologies will likely take too long to be adopted widely enough to effect climate change favorably (Grubler et al., 2016). There is thus an intense need to find ways to accelerate both

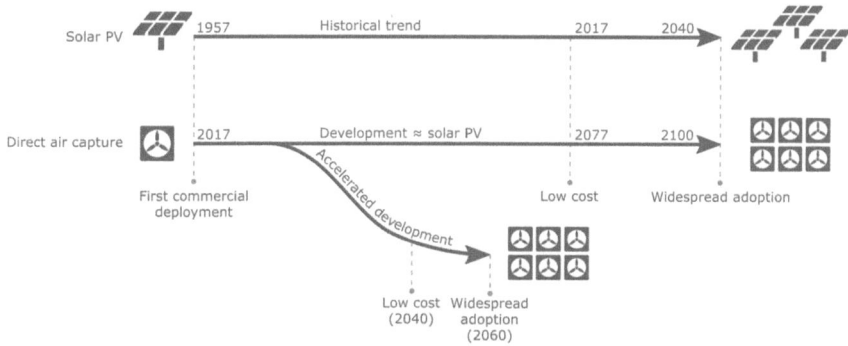

FIGURE 9.1 Comparison of development timeline for PV and direct air capture.

formative phases and the diffusion process for these new technologies. This topic links to an older debate about whether technology deployment is an inherently slow process (O'Neill et al., 2003) or whether it can be accelerated with purposive investment and policy (Hoffert et al., 2003). The urgency of climate change has recently reinvigorated the debate, with similar arguments, albeit with a stronger empirical basis, between those who see possibilities for rapid innovation (Sovacool, 2016; Sovacool and Geels, 2016) and those who point to historical evidence of a more languid pace (Fouquet, 2016; Smil, 2016).

Accelerating innovation for analogous technologies

Given the lengthy time horizons involved in addressing climate change, urgency in technology development and diffusion exists. The relevance of the PV model to low-carbon innovation depends on finding ways to accelerate it. The PV model has plenty of examples of weaknesses in its innovation system that led progress to occur more slowly than necessary. Having looked closely at what has made PV work well, one can consider several ways that would have accelerated PV and would this be relevant to analogous low-carbon technologies.

1. Fund R&D continuously. The boom and bust cycles of R&D investment in PV surely slowed down its progress. It is no surprise that the Japanese took over leadership of PV in the 1980s with a much steadier investment in R&D than the US's cyclical approach. Adequate and steady funding would have avoided the losses associated with knowledge depreciation and move the technology frontier steadily while training scientists and engineers and building absorptive capacity.
2. Early stage procurement. The US Block Buy program from 1975 to 1985 was instrumental in getting companies to develop their production processes towards an industrial process rather than a batch customized one. The US government accounted for a sixth of global PV purchases from 1975 to 1980.

The outcome was not only improved efficiencies but commercial applications of these improvements such that laboratory improvements could be transferred to commercial products more quickly.

3. Trained workforce. Trained scientists and engineers, as well as business people, have been crucial to every country that contributed to inexpensive PV. Some of this training came through formal education, such as at centers of excellence like the University of New South Wales. Others were concentrations of talent such as the National Renewable Energy Laboratory in the US, ISI Fraunhofer in Germany, and the New Energy Development Organization (NEDO) in Japan. Expertise was developed in the private sector through learning by doing, such as at the US entrepreneurial firms in the 1970 as well as Chinese ones in the 2000s.

4. Preserve and codify knowledge. PV's history is one of booms in busts. Each time a breakthrough occurred that launched the industry forward—such as in 1954, 1973, 1994, and 2000—a setback occurred that derailed progress and sent the locus of innovation elsewhere. In each of these bust periods, human capital was squandered as tacit knowledge was lost as people pursued careers outside the PV industry. Sometimes knowledge was proprietary to a firm and was thus kept close even as firms came and went. Codification of knowledge has been important to preserving the value of knowledge, as market volatility is to be expected for other technologies.

5. Disruptive production. Producers of PV adopted a disruptive approach to design and manufacturing multiple times. The first cells for the space program were built with maximum efficiency the goal and reliability a prominent concern. The shift to terrestrial applications during the US Block Buy program in the late 1970s and early 1980s scaled up manufacturing and included relaxation of some of these exacting requirements even as progress in the technology was a thrust of the program. The effort by the Japanese to advance thin film technology in the 1980s also represented a disruptive approach even though the efficiency levels never quite reached what was needed to make them viable. The Chinese compromised on efficiency by about 2 percentage points and were able to wring out huge cost reductions from 2008 to 2012. These cost-reducing efforts were possible because, other than the space program, the reliability of PV was not life-threatening or mission-critical. Producers were able to reduce costs by compromising on product attributes, such as efficiency and reliability, that were less important to adopters than cost.

6. Robust markets. PV served a series of sequential niche markets that generally increased in size and decreased in willingness to pay. PV eventually began to compete with fossil fuel power generation, but only after decades of finding successful niches in satellites, consumer products, and green households. These multiple varied markets provided strong expectations of future markets because the overall demand for PV was robust to adverse changes such as falling energy prices, cancellation of policy support, and saturations of niche

markets. The history of energy implies that prices and markets should be expected to be volatile. Successful technologies will have to find ways to survive during downturns.

7. Knowledge spillover opportunities. PV benefited from technical progress from other industries; the computer industry provided input material, equipment, and production expertise. Wind power, which was more affordable until recently, pioneered and tested new polices and regulatory frameworks. Installer firms benefited from their own experience as well as from that other firms. PV progressed much faster as a result of its ability to assimilate new knowledge from other technologies.

8. Global mobility. So much of the improvement in PV depended on global flows of knowledge. Early on, scientific knowledge was easy to access and allowed rapid global knowledge dissemination. Talent moved quickly to the frontiers of innovation and the largest market opportunities. International visits generated ideas about new applications. Conferences provided a formal mechanism for exchange, but individual and small group visits, such as Martin Green's to the US in 1976 appear to have been just as central. Throughout the history of PV, modules moved inexpensively around the world. Starting in the 1970s, manufacturing equipment moved from the US to Australia, as well as from Germany to the US. Contracts for silicon supply have been international since the 1970s. This enabled specialization while reducing costs. The inexpensive and easy flow of materials, equipment, and finished products has been crucial, despite tariffs that began in 2011. These international flows of ideas, people, and goods enabled countries to focus on the parts of the PV value chain that were most competitive. They also allowed countries to contribute to PV innovation in a way that suited their national innovation system. The inexpensive PV the world found itself with in 2017 is the result of several national innovations working on the problem in parallel. It is hard to imagine a single country stewarding PV successfully from 1954 to present. Competing social priorities always emerge. That these priorities are different across countries and over time makes the global innovation system more effective than any single national one. Facilitating these global knowledge flows is central to accelerating progress of other technologies.

9. Political economy accommodations. Like other new technologies (Weiss and Bonvillian, 2012), all parties will not benefit from the arrival of inexpensive PV. Large entrenched firms and other institutions are threatened by it. These firms have political clout and have successfully sidelined efforts to stimulate PV. The US nuclear industry in the 1970s, Japanese utilities in the 2000s, Spain in 2009, and US utilities in 2014 all set back progress in PV. PV has been most successful when it was able to assemble a supportive advocacy coalition such as in Germany in 1998 and in China in 2010. Other times PV has been successful by accommodating those likely to lose from PV, such as the German FiT that excluded industrial electricity users from the surcharge that paid for the $200 billion program.

All of these characteristics played a role on PV's success, and many of them centrally. All of them could have been enhanced and thus lead to the faster arrival of inexpensive PV. A program to support other low-carbon technologies, which will be necessary to deal with climate change, can use these characteristics and bolster them so that we are not confined to the slow pace of formative phases and diffusion that are a real threat to emerging low-carbon technologies.

References

Bento, N. & Wilson, C. 2016. Measuring the duration of formative phases for energy technologies. *Environmental Innovation and Societal Transitions*, 21, 95–112.

Bento, N., Wilson, C. & Anadon, L. D. 2018. Time to get ready: Conceptualizing the temporal and spatial dynamics of formative phases for energy technologies. *Energy Policy*, 119, 282–293.

Bergek, A., Hekkert, M. & Jacobsson, S. 2008. Functions in innovation systems: A framework for analysing energy system dynamics and identifying goals for system-building activities by entrepreneurs and policy makers. In: Foxon, T. J., Köhler, J.& Oughton, C. (eds.), *Innovation for a Low Carbon Economy: Economic, Institutional and Management Approaches*, Cheltenham, Edward Elgar.

Creutzig, F., Agoston, P., Goldschmidt, J. C., Luderer, G., Nemet, G. & Pietzcker, R. C. 2017. The underestimated potential of solar energy to mitigate climate change. *Nature Energy*, 2, nenergy2017140.

Fouquet, R. 2016. Lessons from energy history for climate policy: Technological change, demand and economic development. *Energy Research & Social Science*, 22, 79–93.

Fuss, S., Lamb, W. F., Callaghan, M. W., Hilaire, J., Creutzig, F., Amann, T., Beringer, T., Garcia, W. D. O., Hartmann, J., Khanna, T., Luderer, G., Nemet, G. F., Rogelj, J., Smith, P., Vicente, J. L. V., Wilcox, J., Dominguez, M. D. M. Z. & Minx, J. C. 2018. Negative emissions—Part 2: Costs, potentials and side effects. *Environmental Research Letters*, 13, 063002.

Gallagher, K. S., Grubler, A., Kuhl, L., Nemet, G. & Wilson, C. 2012. The energy technology innovation system. *Annual Review of Environment and Resources*, 37, 137–162.

Grubler, A., Wilson, C. & Nemet, G. 2016. Apples, oranges, and consistent comparisons of the temporal dynamics of energy transitions. *Energy Research and Social Science*, 22, 18–25.

Hekkert, M. P., Suurs, R. A. A., Negro, S. O., Kuhlmann, S. & Smits, R. 2007. Functions of innovation systems: A new approach for analysing technological change. *Technological Forecasting and Social Change*, 74, 413–432.

Hoffert, M. I., Caldeira, K., Benford, G., Volk, T., Criswell, D. R., Green, C., Herzog, H., Jain, A. K., Kheshgi, H. S., Lackner, K. S., Lewis, J. S., Lightfoot, H. D., Manheimer, W., Mankins, J. C., Mauel, M. E., Perkins, L. J., Schlesinger, M. E., Volk, T. & Wigley, T. M. L. 2003. Planning for future energy resources – Response. *Science*, 300, 582–584.

Nemet, G. F., Callaghan, M. W., Creutzig, F., Fuss, S., Hartmann, J., Hilaire, J., Lamb, W. F., Minx, J. C., Rogers, S. & Smith, P. 2018. Negative emissions—Part 3: Innovation and upscaling. *Environmental Research Letters*, 13, 063003.

O'Neill, B., Grübler, A. & Nakicenovic, N. 2003. Letters to the Editor: Planning for future energy resources. *Science*, 300, 581.

Sivaram, V., Dabiri, J. O. & Hart, D. M. 2018. The need for continued innovation in solar, wind, and energy storage. *Joule*, 2, 1639–1642.

Smil, V. 2016. Examining energy transitions: A dozen insights based on performance. *Energy Research & Social Science*, 22, 194–197.

Sovacool, B. K. 2016. How long will it take? Conceptualizing the temporal dynamics of energy transitions. *Energy Research & Social Science*, 13, 202–215.

Sovacool, B. K. & Geels, F. W. 2016. Further reflections on the temporality of energy transitions: A response to critics. *Energy Research & Social Science*, 22, 232–237.

UNFCCC 2015. *The Paris Agreement. United Framework Convention on Climate Change*.

Weiss, C. & Bonvillian, W. B. 2012. *Structuring an Energy Technology Revolution*, Cambridge, MA, MIT Press.

Wilson, C. 2012. Up-scaling, formative phases, and learning in the historical diffusion of energy technologies. *Energy Policy*, 50, 81–94.

10
APPLYING THE MODEL

We can use what we know about solar's success to inform decisions for other low-carbon technologies that are technologically analogous to solar. The motivation behind this chapter is that we will need many technologies to address climate change, not just solar because a diversity of approaches is likely to enhance effectiveness and minimize adverse consequences. A first obvious comparison to PV is battery storage. But since battery storage technology is already well on its way to scale up, I instead focus on two early stage technologies: direct air carbon capture and sequestration (DAC or DACCS) and small nuclear reactors (SNRs). DAC technology has many characteristics that make it amenable to the PV model, while SNRs (which includes both micro-reactors and small modular reactors, or SMRs) have potential to adopt aspects of the PV model, but their characteristics are less aligned with PVs. After describing each technology, I evaluate the innovation status of each by surveying the landscape of companies and innovation efforts as of late 2018. Using that information, I apply the nine innovation accelerators from Chapter 9: continuous R&D, public procurement, training the workforce, codify knowledge, disruptive production, robust markets, knowledge spillovers, global mobility, and political economy. For each accelerator, I code each technology as high, medium, and low, for both the current status and future potential. Throughout this chapter I focus on the key challenge of scaling up to gigatons worth of removal or of avoided emissions.

Direct air capture technology

Direct air capture with carbon capture and storage provides a way to remove carbon dioxide from the atmosphere via chemical reaction. DAC is one of many negative emissions technologies (NETs) (Minx et al., 2018). NETs are distinct from technologies that reduce greenhouse gas emissions, such as renewables and nuclear power, which can only reduce the flow of carbon dioxide (CO_2) into the atmosphere. Since CO_2 will stay in the atmosphere for a century, removing CO_2 is essential to ambitious climate targets, such as set in the Paris Agreement (Fuss et al., 2018). Other ways of attaining negative emissions include: planting forests, planting mangroves, retaining carbon in soil, and bio-energy with carbon capture and sequestration (BECCS). In a DAC system, a contactor is exposed to ambient CO_2, which is absorbed, for example by reacting it with a nitrogen-based chemical compound, known as an amine. A second reaction then releases the CO_2 from the absorber, so that the CO_2 can be separated, compressed, sent via pipeline to a storage site, and then pumped underground where it will stay. Amine-based DAC was first used in submarines in the 1930s, the space program in the 1960s, and subsequently in indoor air handling applications, in which indoor air needs to be scrubbed of exhaled CO_2. The application of DAC to climate change was first raised in the 1990s (Lackner et al., 1999), however it has been continually dismissed as too expensive (Socolow et al., 2011), in part because of the energy required to concentrate atmospheric CO_2 which is dilute—CO_2 is only 0.04% of air by volume.

Just as in PV, cost reductions have played a central role in DAC innovation efforts and for the most part that effort has consisted of R&D. For example, a 2017 US National Academies report recommended DAC R&D,

> to minimize energy and materials consumption, identify and quantify risks, lower costs, and develop reliable sequestration and monitoring (National Academies, 2017).

As of 2018, R&D activity included: investigating various chemicals for absorbing and adsorbing CO_2 (Kong et al., 2016); minimizing energy use (Lackner, 2013); chemical processes for regenerating solvents (Sanz-Pérez et al., 2016); and dealing with the challenges of working in humid environments (Darunte et al., 2016). Some of this laboratory work has included pilots and prototypes, including for example: a refrigeration-based method for depositing CO_2 (Agee and Orton, 2016), a cooling tower design using hydroxide solution (Holmes et al., 2013), and absorption using an electrolyzed solution (Rau et al., 2013).

Yet nearly all of these R&D efforts do not consider the need to capture at a scale of billions of tons per year. Some work has focused more explicitly on the scale up challenge, including estimating costs of a sodium hydroxide spray system (Stolaroff et al., 2008), cost reductions associated with tens of gigatons of removal per year (Lackner, 2009), macro-economic costs of very large-scale deployment (Pielke, 2009), as well as bottom-up calculations of learning by doing and scale effects (Nemet and Brandt, 2012). A consistent message throughout these efforts is that DAC plants will need to be removing on the order of billions of tons of CO_2 from the atmosphere each year (Hilaire et al., in press). Comprehensive information on costs and potentials can be found in Fuss et al. (2018).

While the magnitude of necessary scale up for DAC to contribute to climate change mitigation is clear, the urgency of that scale up is not well appreciated. A review of modeling scenarios targeting 1.5°C of temperature change found a median rate of CO_2 removal (for all NETs) of 15 Gt CO_2 per year by 2100, with a range of 3–29 (Rogelj et al., 2018). For individual NETs, this amount of removal would imply scaling up to thousands of carbon capture and sequestration (CCS) plants (Herzog, 2011), millions of farms doing soil carbon sequestrations (Woolf et al., in press), or billions of tons of iron added to the ocean for iron fertilization (Boyd and Bressac, 2016). The model scenarios indicate that annual deployment of NETs increases with more ambitious temperature targets, but spans a considerable range (more than an order of magnitude) within each target. The pace of scale up is striking in that even the least stringent targets require adding 150 Mt CO_2 per year of new removal capacity each year between 2030 and 2050 (Nemet et al., 2018a). To put these numbers in perspective, the first large-scale NETs project, a bio-energy with carbon capture and sequestration (BECCS) plant in Decatur, IL USA, will remove about 1 Mt CO_2 per year once in full operation. Worldwide, no other operational projects exceed 0.3 Mt CO_2 per year. These recommended scenarios thus involve bringing hundreds of new plants of Decatur-scale online each year between 2030 and 2050. DAC plants are much smaller—removing around 1000 t CO_2 per year—so would require 100s of thousands of new plants per year.

To further put these results in context: in order to scale up 1 million tons of a specific NET in 2020 to 1 billion tons in 2050, average deployment growth rates of 26% must be sustained for 30 years. Such a scale of growth had been observed for other technologies before, as we have seen for solar PV (Creutzig et al., 2017), but is nonetheless extremely challenging. Consider for example Climeworks, a DAC manufacturer, which at one point had the stated goal to remove 1% of global annual emissions by 2025. Assuming 2025 emissions are similar to the total in 2017, 33 GT CO_2, Climeworks would need to deploy about 350,000 of its DAC-18 units by 2025. I estimate that would require about 1 million shipping containers, which, if shipped by rail as double stacked containers, would fit on a train 5,000 miles long. In line with the company's announcements, assume that two adozen of these units are running by 2020—up from 14 in early 2019. They would have to triple production each year. As shown in Figure 10.1, that rate of scale up (200% per year) is faster than any of the PV companies at their fastest rate of scale up, which include 54% (Sharp), 103% (Jinko), 120% (Suntech), and 129% (Q-Cells). Still, one should not dismiss it as impossible, as it is not far beyond that range. The limiting factor is thus unlikely to be scaling speed but confidence that there will be demand for hundreds of thousands of DAC units per year. Expectations about future markets for air capture are the crucial limiting factor to scale up.

Ultimately, gigaton scale markets for carbon removal will depend on governments putting a price on the damage caused by CO_2 in the atmosphere. That way, operators of DAC would receive a credit or a payment for each ton of CO_2 they remove from the atmosphere. If the prices for these credits are above the marginal cost of removal (mainly heat, electricity, and sorbent replenishment costs), then DAC operators will run their plants. If expected prices are above the average cost

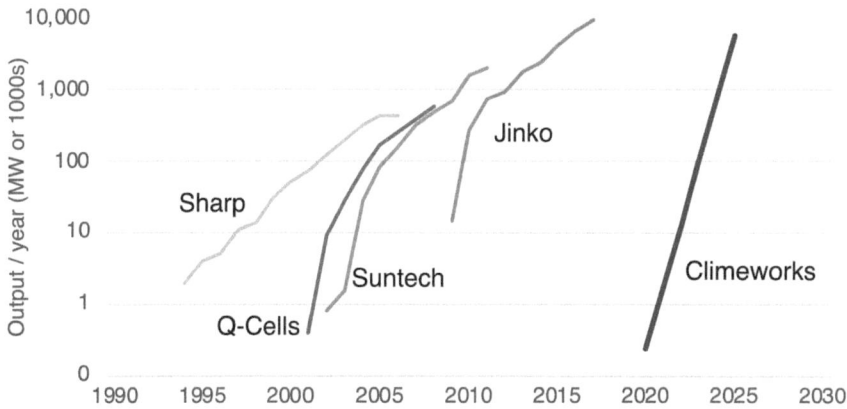

FIGURE 10.1 Comparison of upscaling in PV to upscaling goals for direct air capture.

of removal (including construction costs) then they will build new plants. As of 2018 however, carbon prices were still too low to provide such incentives. Just under half of global emissions have a price of zero and only 9% have a price above $60 per ton (OECD, 2018). Further, almost all of those high-priced emissions are translations of vehicle fuels' taxes into effective CO_2 prices. While these levels could change, as of early-2019 there is almost no policy-created market to compensate DAC for removing CO_2.

However, the lack of adequate carbon pricing does not mean there is no market for DAC. As PV found from 1957 through the 1990s, niche markets can provide a pathway for scale up that can buy time until the mass markets, which provide the bulk of social benefits, materialize (Lackner et al., 2012). DAC firms have found a series of niche markets beginning with submarines and spacecraft to reduce the CO_2 levels of ambient air in closed systems. More generally, high indoor concentrations of CO_2, as much as a factor of ten above ambient levels (Lee et al., 2015), have negative effects on human health, especially for sensitive populations (Kotol et al., 2014). These indoor environments provide an additional niche market for DAC and could increase due to aging populations and sealed building envelopes.

Another set of niche markets involve not just the removal of CO_2 but on the utilization of the concentrated CO_2 that results (Ishimoto et al., 2017). The food industry uses CO_2 to carbonate beverages. Greenhouses increase yields on the order of 20% by increasing the CO_2 concentrations in which plants grow by a factor of two to three. The oil industry injects CO_2 into aging oil fields so that oil, that would otherwise be too difficult to extract, can be recovered. CO_2 can also be utilized to cultivate micro-algae which can be used to produce renewable fuels. Enhanced oil recovery and micro-algae are especially promising niche markets since they do not require highly concentrated CO_2 (Wilcox et al., 2017). To put the size of these niche markets in perspective, food and greenhouse markets are on the order of 20 million tons of CO_2 per year, enhanced oil recovery is about 50 million, while bulk CO_2 removal will require at least 100 times that (Brinckerhoff, 2011). This two-orders-of-magnitude difference need not be discouraging. The largest niche market for PV in 2000 was consumer products, accounting for 15% of PV sales and totaling 40 MW per year (Maycock, 2005). Ten years later the PV market was 21,400 MW per year, a factor of 500 larger than that important niche market.

Carbon pricing on its own has failed to stimulate demand for DAC technologies and there are, generally, no policy instruments that would credit DAC. But despite this, some indications of change are emerging. In the US, the 45-Q tax credit would provide $50 per ton in the form of a credit on federal income taxes. However, it is designed with large power plants in mind and thus contains a lower limit of 100,000 tons per year to prevent large power plants from capturing only 5–10% of their annual emissions. This size limit means that small-scale DAC will be excluded from eligibility for the credit. In California, the low-carbon fuel standard (LCFS) will likely allow DAC to be eligible as an offset of liquid fuel emissions. Those prices are currently well over $100 per ton, however there is large uncertainty about the eligibility of small-scale DAC for the LCFS as its designed for a growing market.

These niche market opportunities—with the potential for a larger policy-driven market at some point—have led a handful of firms to commercialize DAC technology. Most prominently, Climeworks in Switzerland, Global Thermostat in the US, and Carbon Engineering in Canada are all engaged in building commercial devices. Notably all are pure-play companies, i.e. DAC is their only line of business. Others firms, such as Hydrocell, are applying their CO_2 scrubbing technology to DAC applications, but are engaged in other lines of business.

Emerging commercial air capture

Climeworks

Two students at the Swiss Federal Institute of Technology (ETH) in Zurich, Christoph Gebald and Jan Wurzbacher, submitted their first business plan for Climeworks in 2008 as part of an entrepreneurship course at the university. Their concept was to filter CO_2 directly from air and then sell the CO_2. At first, with extremely high oil prices at the time, the most appealing use for the CO_2 was to generate synthetic liquid fuels, and the firm eventually began a partnership with Audi. However, they soon came in contact with Gebrüder Meier Primanatura AG Meier, a family-owned greenhouse business that was interested in a low-cost source of CO_2 to enhance photosynthesis for their cucumbers and tomatoes. The brothers had been purchasing commercial CO_2 that was delivered by truck. They negotiated a contract for Climeworks to supply CO_2 to the Meier Brothers greenhouses in Hinwil, Switzerland near Zurich. The Swiss Federal government funded the early R&D and in 2011 Climeworks closed on $126,000 of incubator financing that would allow them to build the first demonstration plant, which was three orders of magnitude more removal capacity than the laboratory prototypes.

In 2017, the Climeworks Hinwil plant came on-line as the first commercial DAC plant in the world. It uses fans to blow air over a filter made of nano-cellulosic materials in the presence of amines. Within a few hours the 90 square meters of filters become saturated with CO_2, after which the filter is heated with hot water at 100°C to release the CO_2. The plant was installed on top of a municipal waste incineration facility, from which it uses waste heat to desorb the CO_2. This relatively low temperature requirement allows the units to take advantage of low value waste heat that is widely available. The pure CO_2 is then piped 400 meters to the 37,000 square meter greenhouses where it boosts crop yields by 20%. The DAC plant removes about 900 t CO_2 per year, roughly the annual CO_2 emissions of 50 Americans, or alternately of 100 Europeans or Chinese. It also produces about the same amount of fresh water from atmospheric water vapor. To be clear, this CO_2 removal is not permanent as the CO_2 is not sequestered underground at any point. The cost of removing this CO_2 is approximately $600 per ton of CO_2.

Inexpensive manufacturing has been a prominent focus for Climeworks since its inception. In 2014, after closing a $3.4 million Series B round, Climeworks decided to produce a relatively small DAC unit—50 tons removal per year using five square meters of filter. These units can then be shipped with six units to a shipping crate which can then be assembled on site as plants consisting of 1, 3, 18, or 36 units each with the possibility to build plants with thousands of units. This modular scale enables this product to work in a variety of configurations. The Hinwil plant is an 18-unit plant. By 2016, their manufacturing plant could produce 150 DAC units per year, largely made by hand. They have the opportunity to make some compromises on removal efficiency to reduce costs and have continuously have made efforts to remove costs from the production.

In October 2017, Climeworks opened their second DAC plant, "the CarbFix2" project in Hellisheidi, Iceland in partnership with Reykjavik Energy. This smaller facility consists of only one DAC-1 unit capturing 50 tons per year. It uses heat and electricity from Iceland's abundant geothermal resources to run the fans and heat the absorber panel. This plant is different however in that it is the world's first DAC plant to sequester the carbon it removes. The captured CO_2 is dissolved in water and pumped 700m underground where reacts with the basalt bedrock and forms a solid stable mineral deposit.

In August 2018, Climeworks closed on its Series C round with $31 million of funding, mainly from Swiss banks, and in September opened a DAC-3 plant in Troia, Italy. Adjacent to the plant, a solar-powered electrolyzer is producing pure hydrogen from water. In this case, the CO_2 is combined with the hydrogen to make liquified natural gas for use in trucks. This 28-million-euro project is sponsored in part by the European Commission's Horizon 2020 research program.

The diverse end products for these first three commercial projects—agriculture, transportation, and sequestration—reflect Climeworks diverse approach to markets. For CO_2 utilization, in addition to greenhouses, the firm is targeting food and beverages. This market includes CO_2 for carbonated beverages, as an inert gas for packing fresh foods, and as dry ice for refrigeration. At present these markets are served by large industrial gas firms like Linde and Air Liquid. These firms have mature industrial processes with massive economies of scale that dwarf Climeworks' 50 ton per year DAC units by orders of magnitude. But they incur very high costs to transport that gas, mostly by truck, and production sites are highly concentrated—leading some to describe locations as "CO_2 deserts" (Middleton et al., 2014). Whereas CO_2 might sell for $2 per ton next to a manufacturing facility, more remote locations might face costs of over $1000. DAC producers can target these locations as attractive niche markets that allow them to scale up before moving to the larger, but price sensitive, carbon markets. Another set of markets are those for energy, fuels, and materials, for which the Italy plant making liquified natural gas represents a first endeavor. Another pilot project, "Power to X" targets creating a liquid petroleum synthetic fuel by reacting the CO_2 with water and Fischer-Tropsch synthesis. Plastics and base chemicals can also be made with pure CO_2 streams.

Climeworks plans to target these markets over the 2019–2021 period as they scale up their production and reduce their costs; they are not focused on policy-related markets. They claim the technology is already demonstrated as reliable and plan to take out a substantial amount of costs over that time. They have a pipeline of 12–15 new plants through 2021 and a series D round of funding likely in 2020. They had 60 employees in 2018. After 2021, they will be ready for a massive scale up, at which point they will need to be targeting a mass market for DAC rather than niches. As mentioned above, Climeworks had a corporate goal to capture 1% of global emissions by 2025. This will require tripling production of DAC units every year and implies a market for millions of DAC-1 units by then. Climeworks sees an urgency to this scale up that policymakers do not. By 2021 there needs to be a policy regime that credits DAC for removing CO_2 at rates above their costs, which might be in the range of $100–$200 per ton of CO_2. The DAC-18 that came on line in 2017 at Hinwil removes CO_2 at a leveled cost of $600 per ton CO_2, far above that range, but expected for a first-of-a-kind commercial facility. Climeworks claims that they have a pathway to $100 per ton based on removing costs through learning by doing and economies of scale in production. In the long term Climeworks expects to need about four times as much heat energy as electrical energy— and about a third less overall than today's levels. They also see the 2019–2021 period as crucial, their "Valley of Death" (Nemet et al., 2018b), during which they need many customers buying CO_2 from them and after which they need stringent climate policy that will provide credit for DAC removal at above $100 per ton.

Global Thermostat

An American firm, Global Thermostat (GT) is pursuing a similar strategy to Climeworks. Founded in 2010 by Peter Eisenberger and Graciela Chichilnisky, GT was spun out of research at SRI International in Menlo Park, CA where it has been operating pilot and demonstration plants since 2010. The Edgar Bronfman Jr. Family invested $18 million early on, which enabled GT to build its first demonstration plant and develop the technology. They issued debt in 2013, raised $5 million from angel investors in 2015, and won research grants in 2016 and 2017.

The GT process captures CO_2 by exposing air to the contactor, a dry aminopolymer sorbent bonded to a ceramic frame, similar to the arrangement of the catalyst in a catalytic converter in a car. After absorbing CO_2, the contactor is moved to the regeneration box where the CO_2 is then stripped from the sorbent using 85–90°C steam to produce 99% pure CO_2, which can then be used directly or compressed and stored. The contactor then moves back from the regeneration box to the adsorption module where it can begin to adsorb more CO_2. One cycle takes 16 minutes. The process is designed to work in ambient conditions or in environments with much more concentrated CO_2, like industrial flue gas. Each GT DAC module removes 50,000 tons per year and a 40-module array removes 2 million tons—equivalent to the emissions of a coal power plant. GT offers a licensing scheme to make technology accessible to the developing world. GT is looking to similar markets for CO_2 utilization as Climeworks, including: food and beverages, plastics, greenhouses, biofertilizers, industrial gases, synthetic fuels, water desalination, enhanced oil recovery, building materials, as well as the production of algae, and biochar. The existence of CO_2 deserts and the consequent factor of 1,000 difference between the price of CO_2 at the point of production and remote sites creates an opportunity (Middleton et al., 2014). GT has partnerships with two large industrial CO_2 producers, Linde and Air Liquid. Of these niche markets, enhanced oil recovery is the largest.

GT technology has improved dramatically over time. Their first-generation plant built at SRI in 2010 captured 1 kg of CO_2 per hour per square meter of collector area. GT retrofitted that plant in 2012 to double its removal rate. In 2013, their second-generation plant, also built at SRI, could remove 5.6 kg per hour and in 2015 was removing 7.3 kg per hour. GT is finishing work on its first commercial plant in Huntsville, AL where the CO_2 will be used in the beverage industry. It has a removal capacity of 3,000 tons per year (about three times the Climeworks Hinwil plant) and fits in two 40-foot-long shipping containers. It is designed so the next module will be upgraded to 4,000 tons and GT will further adapt this design to develop a unit that will fit in one 40-foot shipping container with a removal capacity of 10,000 tons per year. In addition, it is working on detailed engineering of a larger DAC unit (50 by 6 meters) that will remove 50,000 tons per year.

GT claims the cost of this plant will be $150 per ton of CO_2, which is remarkably low for a first-of-a-kind (FOAK) plant. Such low costs are the result of a gradual process of optimizing the systems since 2010. Capital costs for the Huntsville plant are approximately $2 million. That optimization takes time because it requires building prototypes and operating them in ambient conditions over time to figure out, for example, the timing of cycling and what happens to components after many cycles. GT sees these improvements as indicative of a path to low cost capture, perhaps as low as $50 per ton. They see further cost reductions due to economies of scale, bulk purchasing discounts, and learning by doing—particularly on capital costs. These processes can enable improvements in contactor efficiency, regeneration efficiency, and heat recovery. One of the methods is a disruptive innovation approach. They plan to cycle the absorber after only reaching 30% of its CO_2 absorption capacity. This is less efficient as the sorbent must be cycled every 15 minutes instead of every 60 minutes, but leads to much more CO_2 captured over the course of a day, and thus much lower cost per ton of CO_2. Ultimately, and maybe not too far in the future, they may need a manufacturing partner to produce at very large scale with low margins.

GT is staking a lot on their FOAK plant at Huntsville, anticipating that their second plant will be much easier to sell with a proven installation behind them. One strategy is to be transparent about the performance of that first unit to address skepticism among early adopters. Whether costs will be included in that transparency remains to be seen. They see opportunity in policy with the combination of the 45-Q tax credit and LCFS, a combination that could provide them with reimbursement for carbon removal of $150–200 per ton. But building out capacity will require further investment, even with the confidence of reducing costs through learning by doing and scale. Betting on those cost reductions also entails policy risk in that LCFS price could fall and the 45-Q tax credit could be changed. Niche markets avoid those risks but are limited in size.

Carbon Engineering

Carbon Engineering's (CE) approach is distinct from those of Climeworks and Global Thermostat. Founded in 2009 in Squamish, Canada, CE was based on technology developed by Professor David Keith at Carnegie Mellon University and the University of Calgary. CE built a pilot plant in 2015 that will soon be capturing 1 million tons of CO_2 per year, about half of a coal power plant. CE differs from GT and Climeworks in two important ways. First, rather than target a wide variety of carbon niche markets, CE is focused exclusively on producing low-carbon liquid fuels. Second, rather than manufacturing a small modular device that can be upscaled by arrays of many units, CE is building industrial scale capture plants. As a result, they look more like chemical plants or power plants than those of the other firms.

Their capture plant uses an alkaline hydroxide solution in a cooling tower-like structure to absorb CO_2 and convert it to carbonate. The carbonate solution is then converted into solid pellets of calcium carbonate. The pellets are then heated and the CO_2 is released as a gas. The leftover calcium oxide is then hydrated and used for the initial step. Since starting their plant in 2015, which was generating pure CO_2, CE added fuel synthesis to the plant so that it could use that CO_2 to produce 400 barrels of fuel per year. This so-called "Air to Fuels" technology generates hydrogen by electrolyzing water. It then reacts the hydrogen with the captured CO_2, in the presence of heat and a catalyst, to produce a variety of hydrocarbon fuels, such as gasoline, diesel, and jet fuel.

CE has included high profile investors such as Bill Gates over the course of its $5 million angel investment in 2012, a series A VC round in 2013, a series B round in 2016, and an $11 million convertible loan in June 2018. It will use the $11 million to expand the pilot plant and build its first commercial scale Air to Fuels plant. It plans to use the 2018–21 period to validate the technology at commercial scale so that the risk to new adopters will be minimal. CE's goal is to move to broad adoption from 2021 onwards with individual plants producing 2,000 barrels of fuels per day and removing 1 million tons of CO_2 per year. Their cost goal is $100–150 per ton CO_2, which they document in detail in a publicly available publication (Keith et al., 2018). CE has published substantial data on important aspects of its technology throughout their development (Holmes and Keith, 2012; Holmes et al., 2013; Holmes and Corless, 2014). Their successful optimization of the Air to Fuel process will lead to fuel that costs less than $4 per gallon ($1 per liter). While the focus since 2017 has been on fuels, they also include in their long-term strategy using the captured CO_2 for permanent sequestration, making building materials, and enhanced oil recovery.

Assessment

These three firms are trying to do something similar to what PV is in the process of doing: reduce costs drastically, become deployed massively, and meaningfully address climate change at gigaton scale. Can they follow a similar path to PV? Let's consider evaluate DAC according to the nine innovation accelerators from Chapter 9:

1. *Continuous R&D.* There has been very little public R&D funding on air capture to date. Some individual investors have funded early research in the US, the Swiss government has funded some, the Canadian government has supported pilot work at CE, and in Finland universities are working on DAC devices. This R&D effort is very small considering the scale of the challenge.
2. *Public procurement.* We have not seen any government purchases of DAC devices or purchased CO_2 nor are there any announced plans or proposals to do so.
3. *Trained workforce.* There are on the order of 200 people working for air capture companies worldwide. We see some research in universities but not at sufficient scale to grow the number of trained scientists and engineers involved. These training programs need to expand or companies will need to draw on expertise from other sectors.
4. *Codify knowledge.* Companies are becoming notably more transparent in disclosing the operations, performance, and costs of their plants to date. This information will be helpful for establishing the legitimacy of the industry and spreading knowledge that early plants generate.
5. *Disruptive production.* Small-scale air capture companies, Climeworks and GT are intentionally pursuing a disruptive manufacturing approach. Compromises on removal efficiency have been made by shortening the time involved in the removal-regeneration cycle. This allows for lower cost manufacturing. Costs and components in early prototypes have been removed with each successive generation of the technology.

6. *Robust markets*. Climeworks and Global Thermostat are intending to serve a variety of niche markets rather than focusing on policy dependent market. Their modular scale makes them able to serve a wide variety of applications. Carbon Engineering is focusing on one niche market (fuels) but is also hesitant to serve policy markets directly. Other emerging technologies have seen crashes in market conditions and these three companies seem quite well positioned to survive such an event, due to, for example: falling oil prices, weakened policy stringency, or aggressive pricing by industrial CO_2 producers.
7. *Knowledge spillovers*. Work to date has involved creating machines on a customized basis. There has not been significant use of equipment from other industries that will allow these companies to scale to 100s of Mt of removal capacity per year. Opportunities may exist by using processes from other areas, but that is not clearly occurring as of late 2018.
8. *Global mobility*. Scientists and engineers at these three companies travel internationally, attend conferences, and are visible in international press. Climeworks, in particular, has built its first 14 plants in three different countries. Other than that, we have yet to see much cross-border trade in air capture devices or production equipment. So far, these companies have drawn mainly on domestic knowledge: GT on the US, Climeworks on Switzerland, and CE on Canada—with some of its originating research performed in the US. However, international licensing efforts, proposed at CE and considered at GT, are promising.
9. *Political economy*. DAC seems less in conflict with existing energy producers than renewables. It is being used in the oil industry, and one can imagine hard to decarbonize sectors such as aviation and agriculture using DAC to extend the viability of fossil energy in a carbon constrained world. DAC uses renewables for heat and power. If DAC technology can continue to progress without making enemies of woeful interest groups, its path to scale up will be much steadier than those of its predecessors. It also seems far less likely to encounter public opposition compared to other negative emissions technologies, which are more land intensive.

DAC is achieving moderate performance on four of these nine criteria for accelerating innovation (Table 10.1). At least eight have the potential to be enhanced considerably. None of them look unfeasible. Scaling up to gigatons in time to make a difference on climate change will involve substantial efforts to enhance each of these activity areas. Efforts will have to include support of R&D in the scale up process, the development of specialized production and distribution equipment, and international knowledge flows and licensing. Political economy considerations seem less problematic—although public acceptance of underground CO_2 will be crucial. Active and supportive policy such as public procurement over the next 10 years, stable policy support to enhance scale in the industry, and ultimately markets for carbon credits that are

TABLE 10.1 Status of current effort and future potential of innovation accelerators for direct air capture.

	Accelerators	Current	Potential
1	Continuous R&D	low	high
2	Public procurement	low	high
3	Trained workforce	low	high
4	Codify knowledge	mid	high
5	Disruptive production	mid	high
6	Robust markets	mid	high
7	Knowledge spillovers	low	mid
8	Global mobility	low	high
9	Political economy	mid	high

stringent, credible, and persistent will make a difference in the ability for scale up to proceed at adequate rates.

Small nuclear reactors

Small-modular and micro-scale nuclear reactors are another technology that could be important for climate change and has the potential to learn from the path set by PV. They differ in two important ways from the current worldwide stock of 400 reactors: they are much smaller and they employ a variety of designs rather than the dominant light-water reactor design, which is especially dominant in the US (Allen and Nemet, 2017).

In contrast to the worldwide fleet of nuclear reactors which range in size from 500 to 1500 MW, these new types of reactors aim to provide power at much smaller scales. Small modular reactors (SMRs) have capacities of 50 to 250 MW, about 15% of the size of a full-scale nuclear reactor. Micro-reactors are smaller still, producing 1 to 10 MW of power, about 1% of full-scale size. Moving to such small scales seems counter-intuitive. First, one of the strongest claims from nuclear advocates is that nuclear provides the only large-scale source of low-carbon power, other than hydropower (Bauer et al., 2012; Deutch et al., 2009). SMRs and micro-rectors would appear to be giving up this scale advantage (Cantor and Hewlett, 1988). Second, the history of nuclear power involved a 20-year process, from the late 1950s to the late 1970s, of gradually scaling up plants to the massive sizes that became widespread from the 1980s onwards (Nemet et al., 2018b).

The vision with these smaller models is that they may work in a variety of niche markets for which gigawatt scale is far too large. Their small size implies much larger volumes of manufacturing so that economies of scale can be found in production rather than in unit size (Anadon et al., 2012). Their modular size can allow configuration of almost any size. They can compete in markets that are not being undercut by inexpensive natural gas (Morgan et al., 2018; Roth and Jaramillo,

2017). Their smaller scale might avoid some of the diseconomies of scale that have plagued recent large reactors in the US, Finland, and France with system integration challenges that have led to billions in cost overruns and delays of several years (Abdulla et al., 2013). As such, they may also share characteristics with the PV innovation system. In addition, their pathway from laboratory to demonstrations may also be less capital intensive (Hirth and Steckel, 2016).

The most advanced SMR company is NuScale, a US company designing a modular light water reactor with a capacity of 60 MW. The notion is that it could be configured in plants of up to 12 modules, which would be about half the size of a full-scale plant. The design is meant to be passive, so that instead of using pumps it uses buoyancy to circulate coolant even if the reactor loses power. The smaller units will allow for a standardized manufacturing process in which many units will be produced allowing for economies of scale in production and leaning by doing with iterative improvements to the process. The first NuScale reactor will go online in 2026 at the Idaho National Laboratory. Their plan is to build additional demonstration reactors in six western states, under the Western Initiative for Nuclear (WIN) program. Funding orientated in the US Department of Energy (DOE), which provided more than $200 million in additional funding in 2013 and $40 million in 2018. Investors include Fluor Corporation and Rolls-Royce. NuScale currently employs over 400 people and has raised multiple rounds of venture investment. It has also received stage 1 approval (out of 6) from the US Nuclear Regulatory Commission (NRC).

Another novel nuclear company, although larger than an SMR, TerraPower, also in the US, uses spent fuel from existing large reactors and was founded in 2006 with Bill Gates as an active investor from the outset. TerraPower's Traveling Wave Reactor (TWR) is especially strong in reducing waste by using depleted uranium. The reactor is cooled using molten sodium. It has received funding from the DOE ($40 million in 2016) and has a wide array of partners including companies, universities, and national laboratories. TerraPower currently has 200 employees and their plan is to build a 600 MW demonstration plant by 2018–22 and build full size (1150 MW). Notably, it is building in China where the regulatory processes are easier to navigate. TerraPower plans to develop experience with the technology in China before approaching the US regulatory system.

Transatomic, another US firm developed from technology at MIT, was developing a molten salt reactor with 520 MW capacity. They raised angel investment of $1 million in 2012, and venture capital rounds in 2014, 2015, and 2016. In October 2018, Transatomic acknowledged that they were not able to scale up their technology fast enough and were going to shut down the company. They decided to make their technology know-how open source. Other SMR's include a 225 MW reactor from Westinghouse, a 100 MW System Integrated Modular Advanced Reactor (MSART) from the Korean Atomic Energy Research Institute, a 75 MW pebble-bed high temperature gas cooled reactor from US Department of Energy funded, X-Energy, and a 27 MW reactor from Argentina.

Another set of companies are developing micro-reactors, power plants at scales of 20 MW and less. They are even more oriented toward the PV model because of their small scale and potential fit with a wide variety of niche markets that already exist. One appealing market is that which provides power in remote locations through generators, such as the mining industry. Another market is military installations, which require large amounts of reliable power and, in some cases, need to be mobile. A common challenge is to deal with introducing innovation into a regulatory system designed for large-scale standard reactor designs (Lovering et al., 2018; Ford and Schrag, 2018).

U-Battery, based in Canada is developing a 4 MW gas-cooled reactor, which emerged from Canadian uranium fuels producer, Urenco. It is small enough that it can be installed underground, as a single unit or as a cluster of several 4 MW units. It generates 750°C heat so is intended to be used in locations which can use this heat for industrial processes. In addition to industrial locations, it is intended to serve markets such as back up generation at large nuclear sites, desalination, and hydrogen production.

Oklo, based in Silicon Valley, is building a 15 MW heat-pipe cooled reactor that can run for ten years without refueling. It has been supported by the DOE and has been collaborating with national laboratories. Other micro-reactor ventures include Westinghouse-evinci, a heat pipe reactor that can range from 0.2 to 25 MW of capacity and can run for 10 years without refueling. Holos Generators is developing a 13 MW system that can be transported in a shipping container. They won a $2 million grant from the US Advanced Research Projects Agency for Energy (ARPA-E) in 2018. Ultra-Safe Nuclear Corporation was founded in 2011 and is developing a 10 MW reactor using a fully ceramic micro-encapsulated fuel system. Much of their funding has come from the US space program, NASA, as well as $2 million they received from ARPA-E. Star-Core Nuclear, based in Montreal, is developing a 20 MW high temperature gas reactor and is targeting niche markets in remote sites in Canada. They aim to be operating two pilot plants within five years. Elysium, based in New York, is developing a 20 MW molten chloride salt fast reactor and has received about $3 million in funding from the US Department of Energy.

SMRs and micro-reactors are at a considerably earlier stage than DAC, given no commercial plants have yet been built, nor are they expected to be in the next 3–5 years. Still, we can assess where small nuclear is today, as well as its prospects, with respect to the nine innovation accelerators (Table 10.2).

1. *Continuous R&D.* Levels have declined but R&D has continued and is much stronger than it has been for DAC. R&D might need to be redirected toward innovative designs to reach its potential.
2. *Public procurement.* While no program exists at present, federal power purchase agreements could be applied. Some have called for $2 billion over ten years for four plants (Lovering et al., 2018).
3. *Trained workforce.* Enrollment in nuclear engineering programs has been notoriously volatile (Corradini et al., 2000), but the programs continue to be

TABLE 10.2 Status of current effort and future potential of innovation accelerators for small nuclear reactors.

	Accelerators	Current	Potential
1	Continuous R&D	mid	high
2	Public procurement	low	high
3	Trained workforce	mid	high
4	Codify knowledge	low	mid
5	Disruptive production	low	mid
6	Robust markets	low	mid
7	Knowledge spillovers	low	mid
8	Global mobility	low	mid
9	Political economy	low	mid

productive and the institutional capacity remains strong. Since much capacity exists and training programs are well established, it is reasonable to consider further scale up.

4. *Codify knowledge.* Regulations facilitate codification, since nuclear involves a six-step certification program and some of the results could be disseminated as public information. However, in practice that information has been highly restricted due to concerns about security and claims that information is proprietary. Continued constraints on dissemination are problematic for future scale up.

5. *Disruptive production.* Because of the need to contain radiation within the reactor to avoid accidents and to secure radioactive material throughout the fuel cycle, it is more difficult to take a disruptive innovation approach to nuclear than it is for PV, where compromises on quality yielded large cost reductions.

6. *Robust markets.* The emerging small nuclear industry has identified niches, for example in replacing diesel generators. But it is unclear how large these markets are and also whether other technologies such as battery storage could also compete. It is hard to find substantial niche markets for nuclear that could not be better supplied by diesels or battery storage (Morgan et al., 2018).

7. *Knowledge spillovers.* The most direct concern here is whether the development of small nuclear learns from the operation of large nuclear. Farther afield perhaps are opportunities to learn from other technologies with very high volume but continuing need for high quality. One might consider automobiles where safety is also an issue, volume is high, and scale up has been rapid.

8. *Global mobility.* Mobility of people happens to some extent in nuclear training programs that enroll foreign nationals, as well as employees of multi-national companies. But national security issues limit the extent of collaboration. As for trade and knowledge dissemination, there is the possibility of an international supply chain, but such a prospect is limited by proliferation concerns.

9. *Political economy.* Public acceptance has been very difficult for nuclear, even if it is high among people living near plants. Early efforts to change this perception for small nuclear are underway (Allen and Dines, 2018).

Comparing Table 10.2 to Table 10.1, DAC is in a slightly better position now, and has more potential to follow an accelerated PV model. Small nuclear can also follow the model as it has strong future potential in almost all indicators. However, there are severe challenges to do with limitations on knowledge flows due to proliferation security concerns, the viability of niche markets, limits to the ability to pursue disruptive manufacturing, and public acceptance.

Other models needed

Because energy storage, direct air capture, and small nuclear reactors share characteristics with PV technology, lessons from the history of PV provide useful insights into how to accelerate innovation in these technologies. But other technologies will be needed to address climate change and many of those do not share similarities with PV. Due to the scale of the challenge, its inherently global aspects, and the long time scales involved, the range of technologies is extremely broad (Hawken, 2017). Consequently, a variety of approaches will be needed to support these technologies, accelerate their improvement, and drive their widespread adoption.

If one uses PV as a reference point, one can start to think of climate technologies as falling into one of four categories, or "technology types" (Table 10.3). Type-1 technologies are those we have discussed in this chapter so far. They involve: advanced technology; iterative production processes with many opportunities with which to improve via learning by doing; and have the potential for disruptive manufacturing because opportunities exist for compromising on aspects of the technology to allow substantial reductions in cost. For Type-1 technologies, insights from the history of PV provide a model for accelerating innovation. The PV model might reasonably be applied to a set of other analogous technologies that could have a substantial impact on climate change at scale. These include:

TABLE 10.3 Multiple models for supporting low-carbon innovation.

Technology type	Low-carbon technology target	Innovation model
1. High tech, iterative, disruptive	Direct air capture	Solar PV
2. Low tech, small, distributed	Soil carbon sequestration	Green Revolution
3. Large, system integration intensive	Bioenergy with carbon capture and sequestration	Refineries, chemical plants
4. General purpose	Artificial intelligence	Micro-processors

offshore wind turbines, high altitude wind turbines, fuel cells, desalination, and end use technologies.

Type-2 technologies are too small and low-tech to use the PV model. Examples include planting forests to store carbon in trees and changing agricultural practices to store more carbon in soils. Both of these examples have the potential to remove billions of tons of CO_2 from the atmosphere at relatively modest costs (Fuss et al., 2018). For them, improving the performance or the efficiency of the technology is not the focus. Rather, behavioral change and transaction costs are more likely to be the relevant challenge for scaling up to gigatons of greenhouse gas abatement (Nemet et al., 2018a). Enhancing storage of CO_2 in soils will require widespread adoption of low-till planting, in which fields are not plowed, and cover crops, which retain soil moisture and carbon (Stockmann et al., 2013). These will require the active engagement of millions of farmers around the world to achieve sufficient scale to be substantial for climate change. The technology involved in soil carbon sequestration is so mundane that it's barely worthy of being considered a technology at all. Here we involve Brian Arthur's useful definition of technology as a "means to an end" (Arthur, 2009). The challenge in catalyzing adoption of these agricultural practices is far different from producing electricity from a semi-conductor, iterative improvement by serving sequential niche markets, and economies of production scale. Instead, a more useful model is the Green Revolution (Gaud, 1968), in which substantial increases in crop yields were achieved worldwide by introduction of new techniques—planting high-yield and disease-resistant seed varieties, application of industrially produced fertilizers and chemically produced pesticides, as well as more intensive irrigation (Jain, 2010). Similarly, reforestation and afforestation have the potential to remove billions of tons of CO_2 (Fuss et al., 2018), but their application will involve local land use decisions, particularly working out tradeoffs among competing uses for land including crops, grazing, and CO_2 removal.

Type 3 technologies also have little to learn from the PV model, but not because they are so small like Type-2 but because they are so big. These include low-emission technologies such as full-scale nuclear power, carbon capture at power plants (CCS), low-carbon steel making, and negative emissions technologies such as bio-energy with carbon capture and sequestration (BECCS). These technologies are large, primarily built on-site, and include thousands to millions of parts. System integration is the real challenge, which involves managing complexity, construction costs, and the many entities working together. Some national innovation systems, such as those in China and Korea, are much better equipped to handle these projects than those in the US and Europe. Economies of scale are paramount, implying that building at smaller scales is inefficient and typically unaffordable. Gains from learning by doing are scarce in that far few iterations are available. Consider that the total capacity of nuclear and PV are similar, 400–500 GW. But the entire history of commercial nuclear power has involved building less than 700 reactors whereas the PV industry has built around four billion panels. The improvements that have emerged from billions of iterations are unlikely to apply to the nuclear industry, for which several dozen plants are currently under construction, mostly in China. These large facilities are potentially important for addressing climate change. Indeed, decarbonization scenarios lean heavily on large nuclear and CCS to achieve stringent emissions targets. They will need other types of models. A key challenge for them is to initiate some learning by doing by building a few early plants (Herzog, 2011). How to learn from those plants, manage inevitable failures, and fund the valley of death are central challenges (Nemet et al., 2018b), and quite different from the challenges of accelerating the PV model. Large-scale facilities such as oil refineries and chemical plants are more apt analogues (Remer and Chai, 1990; Rai et al., 2010). Scaling up from small to full size takes time and is a central challenge (Wilson, 2012; Nykvist, 2013).

Finally, Type-4 technologies, while more speculative, may also be important for climate change. "General purpose technologies" (GPTs) are those that have impacts throughout the whole economy, rather than in a particular sector (Ruttan, 2001). Examples include the steam engine, railroads, electricity, the computer, and the internet (Ruttan, 2006; Bresnahan and Trajtenberg, 1995 Freeman, 1994) and in a longer time horizon: use of fire, language, cultivation of plants, and the domestication of animals. Notoriously difficult to predict their emergence and effects, possible candidates for future GPTs include: artificial intelligence, nano-technology, autonomous driving, and synthetic biology. Efforts to accelerate their development would learn from previous GPTs like the computer and the internet. However, GPTs are important in how they are used and diffused throughout the economy not just that they exist. The urgency of addressing climate change implies that the accelerated technology development here will be in speeding the use of emerging GPTs, which might also learn from the network effects so central to exploitation of the internet and the rapid diffusions throughout the economy of computer technology, particularly once they became small and cheap.

Not all technologies can benefit from adopting and accelerating the solar PV model. Small, low-tech Type-2 technologies need to focus on adoption dynamics, risk aversion, and informing millions of micro-decisions. Large, Type-3 technologies need to deal with sharing risk, public financing, codifying knowledge, and tolerating early failures. GPTs, as Type-4 technologies, must learn how to rapidly translate the possibilities created by a new GPT into real technologies that can become adopted quickly and widely. For Type-1 technologies, those similar to PV, there is much potential to learn from the PV model and ways to accelerate the development and adoption processes so that they can affect emissions meaningfully in time to make a difference.

References

Abdulla, A., Azevedo, I. L. & Morgan, M. G. 2013. Expert assessments of the cost of light water small modular reactors. *Proceedings of the National Academy of Sciences*, 110, 9688–9691.

Agee, E. M. & Orton, A. 2016. An initial laboratory prototype experiment for sequestration of atmospheric CO_2. *Journal of Applied Meteorology and Climatology*, 55, 1763–1770.

Allen, T. & Dines, K. 2018. Breaking the nuclear box: What Frank Lloyd Wright can teach us about engaging the community to successfully deploy advanced nuclear technologies, *Third Way*.

Allen, T. & Nemet, G. F. 2017. Energy technology exuberance: How a little humility is good for nuclear, renewables, and society. *Medium*.

Anadon, L. D., Bosetti, V., Bunn, M., Catenacci, M. & Lee, A. 2012. Expert judgments about RD&D and the future of nuclear energy. *Environmental Science & Technology*, 46, 11497–11504.

Arthur, W. B. 2009. *The Nature of Technology: What it is and how it evolves*, New York, Free Press.

Bauer, N., Brecha, R. J. & Luderer, G. 2012. Economics of nuclear power and climate change mitigation policies. *Proceedings of the National Academy of Sciences*, 109, 16805–16810.

Boyd, P. W. & Bressac, M. 2016. Developing a test-bed for robust research governance of geoengineering: The contribution of ocean iron biogeochemistry. *Philosophical Transactions of the Royal Society A: Mathematical Physical and Engineering Sciences*, 374.

Bresnahan, T. F. & Trajtenberg, M. 1995. General purpose technologies 'Engines of growth'? *Journal of Econometrics*, 65, 83–108.

Brinckerhoff, P. 2011. Accelerating the uptake of CCS: Industrial use of captured carbon dioxide. *Global CCS Institute*.

Cantor, R. & Hewlett, J. 1988. The economics of nuclear-power – further evidence on learning, economies of scale, and regulatory effects. *Resources and Energy*, 10, 315–335.

Corradini, M. L., Adams, M. L., Dei, D. E., Isaacs, T., Knoll, G., Miller, W. F. & Rogers, K. C. 2000. *The Future Of University Nuclear Engineering Programs And University Research And Training Reactors*, Washington DC, Dept of Energy, Nuclear Energy Research Advisory Committee (NERAC).

Creutzig, F., Agoston, P., Goldschmidt, J. C., Luderer, G., Nemet, G. & Pietzcker, R. C. 2017. The underestimated potential of solar energy to mitigate climate change. *Nature Energy*, 2, nenergy2017140.

Darunte, L. A., Walton, K. S., Sholl, D. S. & Jones, C. W. 2016. CO_2 capture via adsorption in amine-functionalized sorbents. *Current Opinion in Chemical Engineering*, 12, 82–90.

Deutch, J. M., Forsberg, C. W., Kadak, A. C., Kazimi, M. S., Moniz, E. J. & Parsons, J. E. 2009. *Update of the MIT 2003 future of nuclear power*, Cambridge, MA, Report for Massachusetts Institute of Technology. Retrieved September 17, 2009.

Ford, M. J. & Schrag, D. P. 2018. A tortoise approach for US nuclear research and development. *Nature Energy*, 3, 810–812.

Freeman, C. 1994. The economics of technical change. *Cambridge Journal of Economics*, 18, 463–514.

Fuss, S., Lamb, W. F., Callaghan, M. W., Hilaire, J., Creutzig, F., Amann, T., Beringer, T., Garcia, W. D. O., Hartmann, J., Khanna, T., Luderer, G., Nemet, G. F., Rogelj, J., Smith, P., Vicente, J. L. V., Wilcox, J., Dominguez, M. D. M. Z. & Minx, J. C. 2018. Negative emissions—Part 2: Costs, potentials and side effects. *Environmental Research Letters*, 13, 063002.

Gaud, W. S. 1968. *The Green Revolution: Accomplishments and Apprehensions*. U.S. Agency for International Development

Hawken, P. 2017. *Drawdown: The Most Comprehensive Plan Ever Proposed to Reverse Global Warming*, Penguin.

Herzog, H. J. 2011. Scaling up carbon dioxide capture and storage: From megatons to gigatons. *Energy Economics*, 33, 597–604.

Hilaire, J. R. M., Minx, J., Callaghan, M., Edmonds, J., Luderer, G., Rogelj, J. & Zamora, M. D. M. 2018. Negative emissions and international climate goals – Learning from and about mitigation scenarios. *Environmental Research Letters*, 13, 063001.

Hirth, L. & Steckel, J. C. 2016. The role of capital costs in decarbonizing the electricity sector. *Environmental Research Letters*, 11, 114010.

Holmes, G. & Corless, A. 2014. Direct air capture of CO_2 – an overview of Carbon Engineering's Technology and Pilot Plant Development. *AGU Fall Meeting Abstracts*, 2014.

Holmes, G. & Keith, D. W. 2012. An air–liquid contactor for large-scale capture of CO_2 from air. *Philosophical Transactions of the Royal Society A: Mathematical, Physical and Engineering Sciences*, 370, 4380–4403.

Holmes, G., Nold, K., Walsh, T., Heidel, K., Henderson, M. A., Ritchie, J., Klavins, P., Singh, A. & Keith, D. W. 2013. Outdoor prototype results for direct atmospheric capture of carbon dioxide. *Energy Procedia*, 37, 6079–6095.

Ishimoto, Y., Sugiyama, M., Kato, E., Moriyama, R., Tsuzuki, K. & Kurosawa, A. 2017. *Putting Costs of Direct Air Capture in Context*, FCEA Working Paper Series, 002.

Jain, H. K. 2010. *Green Revolution: History, Impact and Future*, Houston, Studium Press.

Keith, D. W., Holmes, G., St. Angelo, D. & Heidel, K. 2018. A process for capturing CO_2 from the atmosphere. *Joule*.

Kong, Y., Jiang, G. D., Wu, Y., Cui, S. & Shen, X. D. 2016. Amine hybrid aerogel for high-efficiency CO_2 capture: Effect of amine loading and CO_2 concentration. *Chemical Engineering Journal*, 306, 362–368.

Kotol, M., Rode, C., Clausen, G. & Nielsen, T. R. 2014. Indoor environment in bedrooms in 79 Greenlandic households. *Building and Environment*, 81, 29–36.

Lackner, K. S. 2009. Capture of carbon dioxide from ambient air. *European Physical Journal – Special Topics*, 176, 93–106.

Lackner, K. S. 2013. The thermodynamics of direct air capture of carbon dioxide. *Energy*, 50, 38–46.

Lackner, K. S., Grimes, P. & Ziock, H.-J. 1999. Carbon dioxide extraction from air: Is it an option?24th Annual Technical Conference on Coal Utilization and Fuel Systems, March 8–11 1999Clearwater, FL. 885–896.

Lackner, K. S., Brennan, S., Matter, J. R. M., Park, A.-H. A., Wright, A. & Van Der Zwaan, B. 2012. The urgency of the development of CO_2 capture from ambient air. *Proceedings of the National Academy of Sciences*, 109, 13156–13162.

Lee, T. S., Cho, J. H. & Chi, S. H. 2015. Carbon dioxide removal using carbon monolith as electric swing adsorption to improve indoor air quality. *Building and Environment*, 92, 209–221.

Lovering, J., Murray, W., Neeley, J., Nelson, S. & Nordhaus, T. 2018. *Planting the Seeds of a Distributed Nuclear Revolution: The Case for Expedited Licensing and Commercialization of Micronuclear Reactors*, R Street, Clearpath, Breakthrough Institute.

Maycock, P. D. 2005. *PV Technology, Performance, Cost 1995–2010*. Williamsburg, VA, PV Energy Systems.

Middleton, R. S., Clarens, A. F., Liu, X., Bielicki, J. M. & Levine, J. S. 2014. CO_2 deserts: Implications of existing CO_2 supply limitations for carbon management. *Environmental Science & Technology*, 48, 11713–11720.

Minx, J. C., Lamb, W. F., Callaghan, M. W., Fuss, S., Hilaire, J., Creutzig, F., Amann, T., Beringer, T., Garcia, W. D. O., Hartmann, J., Khanna, T., Lenzi, D., Luderer, G., Nemet, G. F., Rogelj, J., Smith, P., Vicente, J. L. V., Wilcox, J. & Dominguez, M. D. M. Z. 2018. Negative emissions: Part 1—research landscape and synthesis. *Environmental Research Letters*, 13, 063001.

Morgan, M. G., Abdulla, A., Ford, M. J. & Rath, M. 2018. US nuclear power: The vanishing low-carbon wedge. *Proceedings of the National Academy of Sciences*, 115, 7184–7189.

National Academies. 2017. *Developing a Research Agenda for Carbon Dioxide Removal and Reliable Sequestration* [Online]. Available: http://nas-sites.org/dels/studies/cdr/ [Accessed 7/31/17].

Nemet, G. F. & Brandt, A. R. 2012. Willingness to pay for a climate backstop: Liquid fuel producers and direct CO_2 air capture. *The Energy Journal*, 33, 53–82.

Nemet, G. F., Callaghan, M. W., Creutzig, F., Fuss, S., Hartmann, J., Hilaire, J., Lamb, W. F., Minx, J. C., Rogers, S. & Smith, P. 2018a. Negative emissions—Part 3: Innovation and upscaling. *Environmental Research Letters*, 13, 063003.

Nemet, G. F., Zipperer, V. & Kraus, M. 2018b. The valley of death, the technology pork barrel, and public support for large demonstration projects. *Energy Policy*, 119, 154–167.

Nykvist, B. 2013. Ten times more difficult: Quantifying the carbon capture and storage challenge. *Energy Policy*, 55, 683–689.

OECD 2018. *Effective Carbon Rates 2018*, Organization of Economic Cooperation and Development.

Pielke, R. A. 2009. An idealized assessment of the economics of air capture of carbon dioxide in mitigation policy. *Environmental Science & Policy*, 12, 216–225.

Rai, V., Victor, D. G. & Thurber, M. C. 2010. Carbon capture and storage at scale: Lessons from the growth of analogous energy technologies. *Energy Policy*, 38, 4089–4098.

Rau, G. H., Carroll, S. A., Bourcier, W. L., Singleton, M. J., Smith, M. M. & Aines, R. D. 2013. Direct electrolytic dissolution of silicate minerals for air CO_2 mitigation and carbon-negative H_2 production. *Proceedings of the National Academy of Sciences*, 110, 10095–10100.

Remer, D. S. & Chai, L. H. 1990. Estimate costs of scaled-up process plants. *Chemical Engineering*, 97, 138–175.

Rogelj, J., Popp, A., Calvin, K., Luderer, G., Emmerling, J., Gernaat, D., Fujimori, S., Strefler, J., Hasegawa, T., Marangoni, G., Krey, V., Kriegler, E., Riahi, K., van Vuuren, D. P., Doelman, J., Drouet, L., Edmonds, J., Fricko, O., Harmsen, M., Havlik, P., Humpenoeder, F., Stehfest, E., & Tavoni, M .2018. Scenarios towards limiting global mean temperature increase below 1.5°C. *Nature Climate Change*, 8, 325–332.

Roth, M. B. & Jaramillo, P. 2017. Going nuclear for climate mitigation: An analysis of the cost effectiveness of preserving existing U.S. nuclear power plants as a carbon avoidance strategy. *Energy*, 131, 67–77.

Ruttan, V. W. 2001. *Technology, Growth, and Development: An Induced Innovation Perspective*, New York, Oxford University Press.

Ruttan, V. W. 2006. *Is War Necessary for Economic Growth? Military Procurement and Technology Development*, Oxford, Oxford University Press.

Sanz-Pérez, E. S., Murdock, C. R., Didas, S. A. & Jones, C. W. 2016. Direct capture of CO_2 from ambient air. *Chemical Reviews*, 116, 11840–11876.

Socolow, R., Desmond, M., Aines, R., Blackstock, J., Bolland, O., Kaarsberg, T., Lewis, N., Mazzotti, M., Pfeffer, A., Sawyer, K., Siirola, J., Smit, B. & Wilcox, J. 2011. *Direct Air Capture of CO_2 with Chemicals: A Technology Assessment for the APS Panel on Public Affairs*, Washington, American Physical Society (APS).

Stockmann, U., Adams, M. A., Crawford, J. W., Field, D. J., Henakaarchchi, N., Jenkins, M., Minasny, B., McBratney, A. B., De Courcelles, V. D., Singh, K., Wheeler, I., Abbott, L., Angers, D. A., Baldock, J., Bird, M., Brookes, P. C., Chenu, C., Jastrow, J. D., Lal, R., Lehmann, J., O'Donnell, A. G., Parton, W. J., Whitehead, D. & Zimmermann, M. 2013. The knowns, known unknowns and unknowns of sequestration of soil organic carbon. *Agriculture Ecosystems & Environment*, 164, 80–99.

Stolaroff, J. K., Keith, D. W. & Lowry, G. V. 2008. Carbon dioxide capture from atmospheric air using sodium hydroxide spray. *Environmental Science and Technology*, 42, 2728–2735.

Wilcox, J., Psarras, P. C. & Liguori, S. 2017. Assessment of reasonable opportunities for direct air capture. *Environmental Research Letters*, 12, 065001.

Wilson, C. 2012. Up-scaling, formative phases, and learning in the historical diffusion of energy technologies. *Energy Policy*, 50, 81–94.

Woolf, D., Lehmann, J., Cowie, A., Cayuela, M. L., Whitman, T. & Sohi, S.in press. Biochar for climate change mitigation: Navigating from science to evidence-based policy. In: Lal, R. (ed.) *Advances in Soil Science: Soil and Climate*, Springer.

11
ACCELERATING INNOVATION

Statistician George Box said that "All models are wrong but some are useful" (Box, 1979). Models are wrong in that they simplify reality and stylize much of the idiosyncrasies and specifics that characterize real world phenomena. Models can be useful when these approximations are informative. When they work well, simplification reveals core driving forces that might be hidden when assessing reality in all its complexity. It is certainly *wrong* to attribute solar's progress to the main drivers of cost reductions: understanding a scientific phenomenon, funding R&D, public procurement, serving niche markets, subsidizing demand, taking advantage of policy windows, and entrepreneurial scale up. The details of policy design played a role, as did Russia's head start in the space race, geo-politics in the Middle East, the 1994 tax law change in China, and the 1998 federal election in Germany. As it has for many technologies, serendipity also affected progress, including chance visits as well as the co-evolution of the semi-conductor industry and the spillovers to PV it provided. Understanding the impact of each of these details has been important to fully assessing solar's trajectory.

But the looming threat of climate instability should make us willing to be wrong here in order to find something useful in PV's success (Xu et al., 2018). The project of decarbonizing the global economy is extraordinarily difficult, an unprecedented task for global environmental governance. It involves meeting the challenge of how societies should address: a truly *global* public goods problem under deep *uncertainty* about future *impacts* and the costs to *address them* with *diverse* perspectives about tolerance for *risk* and *time* preferences, all sustained over the course of several *decades*. The scale of the transformation implies that diversity in approaches will be helpful to mitigate unintended consequences that will inevitably emerge at scale, even for seemingly benign technologies like wind turbines and planting trees. We will need an array of solutions.

The development of inexpensive solar power over the past seven decades represents an impressive achievement of human civilization. Widespread adoption of low-cost solar will help but will not be sufficient on its own—even if it can be sufficient indirectly. I argue in this book that the most important contribution of solar to climate change is in the wrong but useful model it provides for other emerging low-carbon technologies. The solar model is wrong in that it doesn't apply to all low-carbon technologies and because there will always be technological idiosyncrasies that require adaptation of the model. Solar's model is useful in that, as described in Chapter 1, the technology has been successful; its evolution can be accurately traced, as shown in Chapters 2–8; and general findings can be ascertained (Chapter 9) so they can be applied to other analogous technologies (Chapter 10). Having described the evolution of solar and developed general findings, I conclude with three points:

1. We know how to accelerate innovation for technologies analogous to PV,
2. We need to develop other models for different types of technologies, and
3. We need to fit those models in the appropriate national context by considering the distinct attributes of that national innovation system.

Accelerating decarbonization depends on fitting innovation efforts to technologies and national contexts.

Accelerate innovation for PV-like technologies

Low-carbon technologies that are analogous to solar can pursue an accelerated innovation pathway so that they attain meaningful scale in time to meet climate targets. Despite the discussions of the distant future, such as 2050 goals and impacts in 2100, there is an urgency to dealing with climate change that many do not comprehend. Energy infrastructure lasts a long time and CO_2 stays in the atmosphere a long time. Changing the energy system to increase the stability of the climate will take decades. That however, is not a reason to drop climate change to a lower priority issue, just the opposite. Meaningfully addressing climate will require the development and adoption of low-carbon technologies soon enough to avoid the damages associated with further climate change. For example, integrated assessment modeling results show that the key period of deployment for direct air capture is 2030–2050 (Nemet et al., 2018a). For DAC to become widely adopted in that time frame we will need effective, low-cost, reliable, and proven-in-niche-markets air capture by the end of the 2020s. That will require an intense period of innovation between 2020 and 2030.

Understanding the evolution of PV provides lessons for scaling up analogous technologies. There is nothing inevitable about the rapid development and widespread adoption of low-carbon technologies. Rather, intentional policy and purposive investment will be needed and sustained over many years. Government actions can play the role of a catalyst for each of the nine innovation accelerators.

1. *Continuous R&D.* Funding public R&D is the most direct action for this accelerator. Coordinating public R&D across countries, as in Mission Innovation, also plays a role. In addition, governments can create incentives for private R&D and philanthropic funding of R&D, especially in cases in which such incentives are directed at gaps not filled by other mechanisms. The level of funding, targeting funding to address key barriers, and creating incentives for accountability while also encouraging risk taking are central to acceleration.
2. *Public procurement.* Governments can stimulate early stage demand for new technologies by purchasing them. This can take the form of power purchase agreements, holding auctions for set quantities of goods, or guaranteeing prices for specified quantities of a good. These purchases can provide the consistent demand needed for companies to make investments in scale, learning by doing, and worker training.
3. *Trained workforce.* Beyond R&D, public support of education for students and continuing education for needed skills can avoid bottlenecks that emerge as industries scale. In addition, used in conjunction with accelerator #1 on R&D, training scientists and engineers can avoid the crowding out effect that can happen when public R&D is scaled up in an area. By increasing the stock of human talent available, governments can more fully expand the scope of R&D activities, rather than merely shift them from the private to the public sector.
4. *Codify knowledge.* One way governments can play a role in disseminating knowledge gained in innovation activities is by ensuring that at least some of that knowledge is codified, data collected are made publicly available, insights written up in reports, and findings provided at public workshops and presentations. Governments can ensure that this information is made freely available and not concealed as proprietary company information. The history of many technologies shows that industry fortunes come and go and employees are often highly mobile across companies and between industries. Codification can help ensure that knowledge is not lost.
5. *Disruptive production.* Perhaps the item on this list that is least amenable to government activity, there is still a valuable role for creating incentives for companies to creatively design production to make goods that fit what users really value. In PV, certification has been an important way to ensure quality. Firms likely have better information about what attributes should be included in certification schemes, but governments typically have more credibility and thus can use that to promote standards that enable, rather than hinder, disruptive innovation.
6. *Robust markets.* Creating expectations that there will be a market is essential for any low-carbon technology. One must assume that any new technology will be perceived as risky and that almost any government policy supporting it will be vulnerable to change. Consequently, development is often delayed by the existence of a chicken and egg problem in which producers, suppliers,

and customers wait for certainty, which can take a long time to arrive if at all. Governments can shorten that period by helping establish credibility that there will be a market. The German feed-in tariff provided 20-year contracts and gave up substantial flexibility in committing to that long a period. Skewing that trade off toward credibility-enhancing measures at the cost of flexibility is an important way to accelerate innovation by creating expectations that future markets will be robust to changes in markets, technology, and other factors that inevitably will occur.

7. *Knowledge spillovers.* Flows of know-how from one technology to another can be serendipitous. But government can play a role in facilitating them, for example, by funding cross-cutting R&D, convening firms and scientists from disparate backgrounds, and forming industry consortia to facilitate knowledge transfer. Catalyzing these cross-technology knowledge flows is more challenging internationally but can be enhanced with bilateral agreements and international organizations such as the IEA, IAEA, or Mission Innovation.

8. *Global mobility.* Governments are often more focused on reacting to political pressure to impede the flow of people from one country to another rather than facilitate it. However, such facilitation is what is needed to accelerate innovation. International student visas, funding for their travel, work permits for foreign nationals, and international visits by scientists and engineers were all consequential for PV and will be for other low-carbon technologies as well.

9. *Political economy.* For the most controversial of the nine innovation accelerators, it is difficult to be prescriptive because national contexts condition what responses would be most effective. But there is no doubt that large economic actors that stand to lose from the adoption of low-carbon technology will use their political influence to retard the development of their competition. There is ample evidence of these dynamics in PV and all signs point to even stronger application of this power in the future. Acceleration depends on clearing the obstacles that will be deployed. That may include compensating them, actively reducing their influence, or facilitating their evolution to firms that can take advantage of the opportunities provided by better low-carbon technology.

Governments can both enhance these accelerators and mitigate the threats to these accelerators and the global climate innovation system they would support.

Fit technologies to the right innovation model

The scale of the climate challenge requires massive adoption of low-carbon technologies beyond those that are analogous to solar. Consequently, we will need other innovation models, as well as other accelerators. Under the premise that different innovation models work best for different technology types, Chapter 10 laid out four broad classes of technologies relevant to climate change. Type-1 is the reference case, those technologies that are analogous to PV. Type-2 technologies are smaller and more low-tech than PV. Type-3 technologies involve larger unit scale than PV. Type-4 technologies are less directly comparable because they are general purpose rather than a distinct technology application—that is they can make other applied technologies feasible or perform better. We can thus think of a matrix of four low-carbon technology models (Figure 11.1) with a potential analogy for each that can guide future development of the technologies we will need for low-carbon technology. Governments can play a role in accelerating innovation in each.

First, Type-1 is the PV reference group: technologies that are high technology, massively iterative, and are amenable to disruptive innovation attributes. These technologies have characteristics that make them similar to PV in the responsiveness to the types of drivers that led to low-cost PV. I summarize those drivers in Chapter 10 and in more detail in Chapter 2.

Second, Type-2 technologies are low-tech, involve hyper-distributed adoption processes, and require know-how, but are only partly amenable to infusions of R&D and technology spillovers. There are only limited opportunities to industrialize or automate the processes. Examples of low-carbon technology in this category include soil carbon sequestration, reforestation and afforestation, as well as the production and application of biochar (Griscom et al., 2017). Successful examples of previous analogs include hybrid seeds and the Green Revolution (Gaud, 1968). Avenues to improve and accelerate adoption for Type-2 are quite different than those for Type-1 PV (Ruttan, 1996). They involve understanding local adoption dynamics, evolving community acceptance of new approaches, and communicating benefits and risks to a diverse set of stakeholders (Trevisan et al., 2016). In contrast to Type-1 PV-like technologies, speeding up Type-2 innovation requires a diverse set of applications in heterogeneous adoption settings.

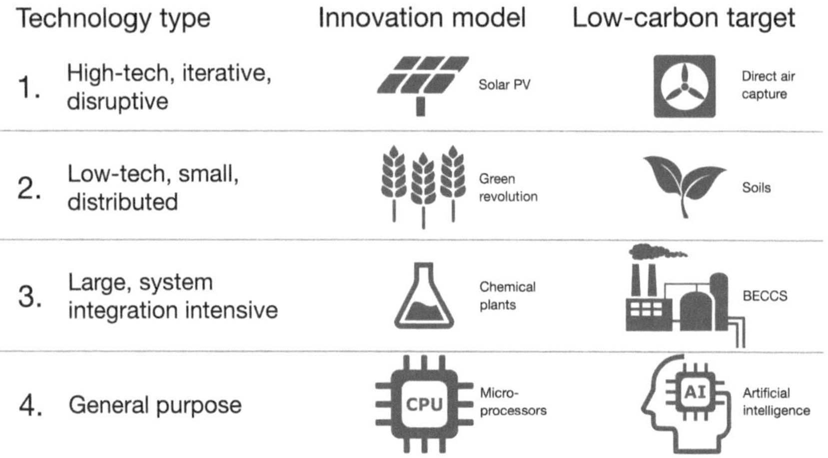

	Technology type	Innovation model	Low-carbon target
1.	High-tech, iterative, disruptive	Solar PV	Direct air capture
2.	Low-tech, small, distributed	Green revolution	Soils
3.	Large, system integration intensive	Chemical plants	BECCS
4.	General purpose	Microprocessors	Artificial intelligence

FIGURE 11.1 Four innovation models for four technology types.

Third, Type-3 technologies are large, with units of massive scale rather than large manufacturing scale. They are system integration intensive; substantial science and technology is involved; and a relatively low number of installations limit iterative improvement (Nemet et al., 2018b). Examples of low-carbon technology in this category include bio-energy with carbon capture and sequestration (BECCS), bio-refineries, carbon capture at fossil fuel power plants, and low-carbon steel production via hydrogen reduction and electric arc furnaces (Lechtenböhmer et al., 2016). Successful past examples of these types of technologies include oil refineries and chemical plants (Merrow et al., 1981). Accelerating innovation in Type-3 technologies is also likely to be quite different than for Type-1 in that it includes funding demonstrations, building strong organizational capability and developing system integration expertise, all at the large unit scales inherent in Type-3 (Åhman et al., 2018).

Fourth, Type-4 technologies are general purpose technologies (GPTs). These technologies tend to involve scientific breakthroughs; they open the door for an array of new applications and enable new technologies (Helpman, 1998; Mowery and Simcoe, 2002). Examples in the past include the steam engine, micro-processors, and the internet (Pearson and Foxon, 2012). Because they involve general applications rather than specific technologies, one would not classify them as specifically low-carbon. However, examples of GPTs with the potential to open up new possibilities for low-carbon technologies include artificial intelligence, robotics, and synthetic biology. How to stimulate GPTs has always been somewhat murky, as they sometimes seem to simply appear (Ardito et al., 2016). But the conditions for their emergence can also be supported by government, and probably must be since they almost always involved substantial spillovers rather than only sustained corporate investment (Mazzucato, 2013; Mazzucato and Semieniuk, 2017). Companies thus have only weak incentives to invest in developing them, even if they have strong incentive to later adopt them. Government can support GPTs though funding science and education as well as more applied R&D. Early support of GPTs can involve finding early applications for them, which can strengthen incentives to develop them. Global dissemination of them, through scientific exchange and licensing can also hasten their development and use.

These sketches of policy directions are intended to make the case that a variety of avenues exists to accelerate innovation in low-carbon technologies—not just for Type-1 PV analogues, but for much different types of technologies as well. The key research needs for Type-2, Type-3, and Type-4 technologies are to develop effective and affordable innovation accelerators for them. But those accelerators will need to fit into the right context as well.

Fit innovation models to the right national innovation system

Distinct innovation capabilities in each country mean that some of these models are better suited for some innovation environments over others. The combinations of the inherently global aspects of low-carbon innovation, and insights from the National Innovation System (NIS) literature, makes clear that innovation models will function well in places with the proper institutions and capacity to make them function well.

From a single country's perspective, this also implies that some types of technologies will be more suited to it than others. This comparative advantage is quite obvious in considering energy resources. Windy Denmark leads the world in wind turbine technology and resource-rich China builds the most efficient coal power plants in the world. But for future low-carbon technologies, a country's NIS is also likely to influence the direction of low-carbon innovation. For example, in viewing the US, one can consider the types of innovations that have been successful over the past couple of decades. Knowledge intensive, small unit size, massive production volumes, global supply chains, and disruptive approaches appear to characterize much of what has worked—increasingly with a government role (Block and Keller, 2008). Even if most of the production of Type-1 technologies like smart phones and solar panels has moved offshore, much of the invention and design behind them, as well as substantial value added, is situated in the US. The US has also been a strong source of Type-4 general purpose technologies, such as the internet. The US innovation system has been characterized on the technology supply side by a robust and diverse research enterprise with high levels of funding, while on the demand side it has a large domestic market that is relatively competitive and thus facilitates firm entry (Mowery and Simcoe, 2002). Recently the US appears to be struggling with large-scale Type-3 technologies. The large-scale nuclear renaissance appears to have mostly passed the US by, with two large reactor projects abandoned in 2017, amid delays and massive cost overruns, with only two more under construction as of early 2019. Plans for large-scale carbon capture have also struggled with similar issues in delays, excessive costs, and cancellations. In contrast, Type-3 technologies, large-scale projects, seem much better suited to what has been successful in China, with its expertise in "organizational capacity" described in Chapter 7. China is building over sixty nuclear power plants. Even more impressively, its two decades of massive infrastructure development, now extending externally via the Belt and Road Initiative, make it far more likely to be capable of executing on large system-integration intensive Type-3 technologies. China is making efforts to excel at Type-1 technologies through R&D and technology upgrades, linked to its manufacturing prowess (Ball et al., 2017). It is clearly targeting Type-4 innovation with its massive investment in artificial intelligence (Barton et al., 2017) while its large population and data collection capabilities position it for rapid machine learning (Larson, 2018).

It is somewhat murkier how the US and China fare with small-scale Type-2 technologies. US farmers have traditionally been rapid adopters of new technology, whether hybrid seeds in the 1930s or genetically modified organisms in the 1990s and 2000s. However, agriculture in the US is increasingly characterized by consolidation and it is unclear whether this trend will on balance encourage the adoption of new low-carbon practices, because of large resources available, or discourage them, due to near monopsony and monopoly power. China's agricultural sector is amenable to Type-2 innovation, although the intense focus on food security will likely eliminate considerations of Type-2 technologies with adverse side-effects on the reliability of food production.

Of course, the two countries mentioned here only account for less than a quarter of the world's population. Country-level decisions about which types of low-carbon innovation they will pursue, if any, will affect the global patterns of low-carbon innovation. We should expect them to be diverse and, to some extent, difficult to predict. Coordinating them seems unlikely—even if some type of monitoring to identify gaps and emerging opportunities has potential for large social benefits (Sanchez and Sivaram, 2017). In short, we need different innovation models for different technologies—but we also need to fit the right innovation models with a country's distinct national innovation system.

Eight reasons for optimism

I am convinced that even though transforming the world's energy system to better meet people's needs will be hard, there are many reasons to be optimistic. In this book and in teaching courses on energy and the environment, I emphasize core characteristics of the problems that make them difficult to solve. For example, even though most people would like an energy system that is cheap, clean, and reliable, they argue fiercely about which of the three they believe is most important. Transitioning to a new energy system with fewer tradeoffs among these three goals would help make energy issues less contentious. But as discussed in Chapter 1, that change is likely to be slow because energy infrastructure—power plants and transmission lines—lasts many decades. It is even harder to make a quick impact on climate change since greenhouse gases linger in the atmosphere for a century or more. Note that this longevity differs from more familiar air pollutants, like sulfur dioxide and nitrogen oxides, for which rain and deposition can remove much of their human impact over days or weeks. Climate change is further complicated by persistent uncertainty about where, when, and how intense the damages will be—even if the evidence of changes and of the human role is clear.

Just under a decade ago, following the disappointment of the December 2009 UN Copenhagen Climate Change Conference in Copenhagen, I began to realize that I was making these points about the difficulty of the problem to my students a little too well. Many students emerged from my courses discouraged from working on climate change issues. That was the last impact I wanted to have. It was also not how I personally felt about the problem. I have many colleagues who have devoted their careers to working on climate issues and I knew that they were not doing so just because it was worth trying. I think people actually think we can successfully address climate change. And so I began seriously asking myself why I continued to work on climate change. I took notes, grouped the ideas, and began ending my courses with a discussion of this list of reasons to be optimistic that we can address climate change.

The first reason provides the central motivation for this book project. Even though politicians, governments, and the international community have been dithering on climate policy for 30 years, low-carbon technologies have continued to progress, improve, and become cheaper. As we have seen, solar prices are low and continue to fall. But so do prices for electric vehicles, wind turbines, and integrating information technology into our energy system. The accumulation of improvements is making the transition to a cheap, clean, and reliable energy system far more feasible than it was just a few years ago.

Second, international cooperation on climate change is stronger than it has been since the Kyoto Protocol was signed in 1997. Just over three years ago, the Paris Agreement on Climate Change did not exist. Now nearly 200 countries have signed on. That agreement, while insufficient, is crucial to making progress. The bottom-up process of nationally determined contributions has brought a wide array of countries on board and has allowed them to customize their plans to their own national situations. The five-year review process provides a built-in mechanism to enhance ambition, and the agreements on transparency and reporting have introduced new accountability. While the national plans to date do not put us on a pathway to avoiding temperatures above 2°C, we are in far better shape with the Paris Agreement in place than without.

Third, as we have seen in PV, countries and local governments are learning from the experience of other governments in climate-related policy more generally. They take ideas from each other. Over two decades of diverse policies and initiatives have now been tried: research funding, tax credits, demonstration projects, feed-in tariffs, renewable portfolio standards, building codes, cap and trade systems, and hybrid policies. We have seen policies implemented by municipalities, provinces, national governments, and at the international level. Evaluations about what works are far more available than they were just a few years ago. The opportunities for policy learning in climate change are higher than they have ever been. As we have seen in PV, policymakers have often done well when adopting proven policy ideas from other jurisdictions. Policy diffusion is as important as technology diffusion and it can also happen quickly.

Fourth, we have precedents from other areas of environmental concern. The determination to address a problem has overcome challenges involved with international cooperation, long time-horizons, and powerful affected interests. The phase out of lead in gasoline, reduction of acid rain, and repairing the ozone hole are successes we can learn from. None are quite as difficult as climate change, but like the model of PV for technology, it is much easier to learn from small successes than from large failures.

Fifth, communities have strong incentives to adapt to a changing climate. They will need to manage the impacts of more severe weather, rising sea levels, and shifting growing seasons, among many other changes. Communities know they cannot wait for others to go first in order to protect themselves from the inevitable changes to come. In contrast to reducing emissions, investing in adaptation brings forth climate related benefits that are near term and immediate. Innovation in adaptation is not only possible, but its development, through successful adaptation models, will also enhance communities' abilities to make them more resilient to climatic changes.

Sixth, substantial co-benefits result from reducing greenhouse gas emissions. Transforming to low-carbon energy production and use provides near-term local benefits, such as cleaner air, energy cost savings, energy security, and increased employment. In locations with rapidly developing economies, these co-benefits may comprise the primary motivation for an energy transition, with climate change as a secondary consideration. That these co-benefits are near-term and local can make them more politically salient than climate.

Seventh, low-carbon economies need not involve sacrifice. One can observe an increase in examples of cities and countries with much less energy-intensive economies, but high quality of life—for example in terms of health, education, and income. Further, one should find encouragement in noting that, even though policy laggards like the US federal government are falling behind, there are other places that are developing the technologies of the future. This is surely a missed opportunity for the US, but the world would be far worse off if China, as well as countries throughout Asia, the Middle East, and Europe were not actively investing.

Finally, people in their 20s and 30s have high stakes in the outcomes of what society does on climate change. As a teacher, it is heartening to see their engagement, whether as activists, technologists, or in an array of roles that they are inventing for themselves.

Over time this list has lengthened and each item has become more convincing. One could write several pages or more substantiating each point. Lately, I have been encouraged by the feeling that I am missing other reasons that may be very important motivators to others. Addressing climate change is a daunting challenge and precious time has been squandered, but many good things are happening and we are moving in the right direction. Of all of these reasons, I find the first one the most compelling. As shown in this book, there is a mountain of evidence supporting the claim that PV is getting better. Some of the same innovation dynamics are happening in other important technologies, especially energy storage and digitalization. Improvements in climate technology also bolster the credibility of the other reasons. For example, technology supports the second reason, collective action, by facilitating international agreement and making low-carbon economies compatible with high quality of life. Positive interactions among these reasons provide further encouragement. Still, the emergence of an increasingly favorable set of conditions does not detract from the most important implication for climate policy from this project—we need to move faster.

References

Åhman, M., Skjærseth, J. B. & Eikeland, P. O. 2018. Demonstrating climate mitigation technologies: An early assessment of the NER 300 programme. *Energy Policy*, 117, 100–107.

Ardito, L., Petruzzelli, A. M. & Albino, V. 2016. Investigating the antecedents of general purpose technologies: A patent perspective in the green energy field. *Journal of Engineering and Technology Management*, 39, 81–100.

Ball, J., Reicher, D., Sun, X. & Pollock, C. 2017. *The New Solar System: China's Evolving Solar Industry And Its Implications for Competitive Solar Power In the United States and the World*, Stanford, Steyer-Taylor Center for Energy Policy and Finance.

Barton, D., Woetzel, J., Seong, J. & Tian, Q. 2017. Artificial intelligence: Implications for China, New York, McKinsey Global Institute.

Block, F. & Keller, M. 2008. *Where Do Innovations Come From? Transformations in the U.S. National Innovation System, 1970–2006*, Washington, The Information Technology and Innovation Foundation, .

Box, G. 1979. Robustness in the strategy of scientific model building. In: Launer, R. L. & Wilkinson, G. N. (eds.), *Robustness in Statistics*, Elsevier.

Gaud, W. S. 1968. *The Green Revolution: Accomplishments and Apprehensions*, U.S. Agency for International Development

Griscom, B. W., Adams, J., Ellis, P. W., Houghton, R. A., Lomax, G., Miteva, D. A., Schlesinger, W. H., Shoch, D., Siikamäki, J. V., Smith, P., Woodbury, P., Zganjar, C., Blackman, A., Campari, J., Conant, R. T., Delgado, C., Elias, P., Gopalakrishna, T., Hamsik, M. R., Herrero, M., Kiesecker, J., Landis, E., Laestadius, L., Leavitt, S. M., Minnemeyer, S., Polasky, S., Potapov, P., Putz, F. E., Sanderman, J., Silvius, M., Wollenberg, E. & Fargione, J. 2017. Natural climate solutions. *Proceedings of the National Academy of Sciences*, 114, 11645–11650.

Helpman, E. 1998. *General Purpose Technologies and Economic Growth*, Cambridge, MA, MIT Press.

Larson, C. 2018. China's massive investment in artificial intelligence has an insidious downside. *Science*, 8.

Lechtenböhmer, S., Nilsson, L. J., Åhman, M. & Schneider, C. 2016. Decarbonising the energy intensive basic materials industry through electrification – Implications for future EU electricity demand. *Energy*, 115, 1623–1631.

Mazzucato, M. 2013. *The Entrepreneurial State: Debunking Public vs. Private Sector Myths*, London and New York, Anthem Press.

Mazzucato, M. & Semieniuk, G. 2017. Public financing of innovation: New questions. *Oxford Review of Economic Policy*, 33, 24–48.

Merrow, E. W., Phillips, K. & Myers, C. W. 1981. *Understanding Cost Growth and Performance Shortfalls in Pioneer Process Plants*, Santa Monica, CA, The RAND Corporation.

Mowery, D. C. & Simcoe, T. 2002. Is the Internet a US invention?—An economic and technological history of computer networking. *Research Policy*, 31, 1369–1387.

Nemet, G. F., Callaghan, M. W., Creutzig, F., Fuss, S., Hartmann, J., Hilaire, J., Lamb, W. F., Minx, J. C., Rogers, S. & Smith, P. 2018a. Negative emissions—Part 3: Innovation and upscaling. Environmental Research Letters, 13, 063003.

Nemet, G. F., Zipperer, V. & Kraus, M. 2018b. The valley of death, the technology pork barrel, and public support for large demonstration projects. *Energy Policy*, 119, 154–167.

Pearson, P. J. G. & Foxon, T. J. 2012. A low carbon industrial revolution? Insights and challenges from past technological and economic transformations. *Energy Policy*, 50, 117–127.

Ruttan, V. W. 1996. Induced innovation and path dependence: A reassessment with respect to agricultural development and the environment. *Technological Forecasting and Social Change*, 53, 41–59.

Sanchez, D. L. & Sivaram, V. 2017. Saving innovative climate and energy research: Four recommendations for Mission Innovation. *Energy Research & Social Science*, 29, 123–126.

Trevisan, A. C. D., Schmitt-Filho, A. L., Farley, J., Fantini, A. C. & Longo, C. 2016. Farmer perceptions, policy and reforestation in Santa Catarina, Brazil. *Ecological Economics*, 130, 53–63.

Xu, Y., Ramanathan, V. & Victor, D. G. 2018. Global warming will happen faster than we think. *Nature*, 564, 30–32.

END MATTER

Interviews or Informative Discussions were conducted with the following individuals.

Last name	First name	Country	Affiliation
Ahman	Max	Sweden	Lund University
Allen	Todd	US	University of Michigan
Arent	Doug	US	National Renewable Energy Laboratory
Arjmand	Mehrdad	US	NovoMoto
Beutller	Christoph	Switzerland	ClimeWorks
Breyer	Christian	Finland	Lappeenranta University of Technology
Corradini	Mike	US	University of Wisconsin
Cronje	Christian	South Africa	Juwi Renewable Energies
Cuevas	Carolina	Chile	Fondacion Chile
Dufey	Annie	Chile	Ministry of Energy
Fath	Peter	Germany	RCT Solutions
Fell	Hans-Josef	Germany	EnergyWatch
Gallagher	Kelly	US	Tufts University
Gay	Charlie	US	Department of Energy

Last name	First name	Country	Affiliation
Geilen	Dolf	Abu Dhabi	International Renewable Energy Agency
Gipe	Paul	US	Wind Works
Goldschmidt	Jan Christoph	Germany	Fraunhofer Institute for Solar Energy Systems
Green	Martin	Australia	University of New South Wales
Grubler	Arnulf	Austria US	International Institute for Applied Systems Analysis
Haverkamp	Helge	Germany	Schmid Group
Hawkey	Neville	South Africa	Independent contractor
Hu	Gao	China	China National Renewable Energy Center
Jiayang	Wang	Japan	University of Tokyo
Kaizuka	Izumi	Japan	RTS Corporation
Khan	Nuaman	Pakistan	Grace Solar
Kimura	Osamu	Japan	Central Research Institute of Electric Power Industry
Kurtz	Sarah	US	University of California
Laird	Frank	US	University of Denver
Li	Junfeng	China	National Climate Strategy Center of NDRC
Liu	Yifeng	China	Sunlectric
Lossen	Jan	Germany	ISC Konstanz
Margolis	Robert	US	National Renewable Energy Laboratory
Martinot	Eric	Japan	Institute for Sustainable Energy Policies
Marukawa	Tomoo	Japan	University of Tokyo
Maycock	Paul	US	Department of Energy

Last name	First name	Country	Affiliation
Neuhoff	Karsten	Germany	German Institute for Economic Research (DIW)
Nillson	Lars	Sweden	Lund University
Noel	Lance	Denmark	Aarhus University
O'Connell	Ric	China	Gridlab
Peng	Peng	China	Chinese Renewable Energy Industries Association
Perlin	John	US	University of California
Pflueger	Antonio	Germany	German Federal Ministry for Economic Affairs and Energy
Ping	Eric	US	Global Thermostat
Qi	Ye	China	Tsinghua University
Rai	Varun	US	University of Texas
Rein	Alex	US	Prairie Light Solar
Shi	Zhengrong	China	Suntech
Shimamoto	Minoru	Japan	Hitotsubashi University
Sivaram	Varun	India	ReNew Power
Splinter	Mike	US	Applied Materials
Su	Jun	China	Tsinghua University
Sugiyama	Masahiro	Japan	University of Tokyo
Sugiyama	Taishi	Japan	Canon Institute for Global Studies
Suzuki	Akio	Japan	Sharp
Swanson	Dick	US	SunPower
Szpitalak	Ted	China	Pan Asia Solar
Urey	Emmanuele	Liberia	Liberia Engineering & Geo-Tech Consultants
Wang	Jinzhao	China	Development Research Center of State Council
Wang	Nan	Japan	Research Institute of Innovative Technology for the Earth

Last name	First name	Country	Affiliation
Watanabe	Chihiro	Japan	Tokyo Institute of Technology
Weber	Eicke	Singapore	University of California
Werner	Tom	US	SunPower
Wilson	Charlie	UK	University of East Anglia
Wilson	Greg	US	National Renewable Energy Laboratory
Wiser	Ryan	US	Lawrence Berkeley National Laboratory
Wolfe	Philip	UK	Former CEO BP Solar
Yamaguchi	Masafumi	Japan	New Energy and Industrial Technology Development Organization
Zhou	Dequon	China	Nanjing University
Zhou	Peng	China	China University of Petroleum
Anonymous		Saudi Arabia	King Abdullah University of Science and Technology
Anonymous, 1		Saudi Arabia	Saudi Aramco
Anonymous, 2		Saudi Arabia	Saudi Aramco
Anonymous		South Africa	South African Photovoltaic Industry Associatio

INDEX

Note: page numbers in italic type refer to Figures; those in bold type refer to Tables.

1968 student protests, Germany 106, 108–109, 113, 118

a-Si (amorphous silicon) 94, 95–96, 97, 103
Aachen, Germany 76, 111, 161, 170
Abu Dhabi 9
acid rain 110, 220
Actis capital 148
Adams, William Grylls 56
adoption 183; early adopters 31, 36, 55; local processes 159, 161–163; peer effects in 160, 162, 183; US military adoption 24, 36, 55, 61, 63
Advanced Research Projects Agency (ARPA-E), US 203
ADWEC 9
AEG 109
affordability of energy 2, 3, 4
afforestation 191, 206, 216
Africa 107, 169
Agency for Industrial Science and Technology (AIST), Japan 85, 90, 91, 92, 99
Agora Energiewende 107
Aihua Wang 140
Air Liquide 195
air pollution: cleanliness of energy supply 2, 3, 4; as solar PV adoption reason 162
Air to Fuels technology 198–199
AIST (Agency for Industrial Science and Technology), Japan 85, 90, 91, 92, 99

Alpha Real 98, 111
Amoco 79
Amtech 149
analogous technologies 190–207; accelerating innovation for 20, 185–188, 212–215; right innovation models for 215–216, *216*; right NIS (National Innovation System) for 217–219; technology types 205–207, **207**
Annan, Richard 74
anti-nuclear movement, 1970s, Germany 106, 108–109
Apollo Project 66, 68, 71
Applied Materials 123, 126, 151, 153
Applied Solar Energy: Angewandte Solarenergie (ASE) 111, 124
Arab Oil Embargo 1973 3, 39, 65, 67–69, 79, 90, 91, 109, 113, 178
Arabian Peninsula 128
ArcoSolar 79, 80
Argentina 202
ARPA-E (Advanced Research Projects Agency), US 203
Arthur, Brian 57, 206
Asbeck, Frank 126, 146
ASE (Applied Solar Energy: Angewandte Solarenergie) 111, 124
ASE/RWE-Schott Solar 112
Asia 107
Assembly Bill 1575, California 76–77
Association for Applied Solar Energy 61

Astropower 146
Asys 123
Atomic Energy Research Institute, Korea 202
ATS 149
Australia 2, 40, 98, 178; and China 133, 134, 135, 139–141; *see also* UNSW (University of New South Wales)
Automatic Power 62

Baccini 123
Baden-Würtemberg, Germany 40
"balance of system" 40, 159; *see also* soft costs
battery storage 20, 168, 169, 190
BBC 109
BECCS (bio-energy with carbon capture and sequestration) 191, 206, 217
Becquerel, Edmund 20, 35, 55, 56, 60, 177
Beijing Solar Research Institute 149
Beijing Sunlectric Technology Company 156–157
Bell, Emerson 60
Bell Labs 20, 30, 35, 47, 55, 56, 57–61, *60*, 62, 74, 89, 109, 178, 183, 184
Belt and Road Initiative, China 218
Berman, Elliot 62, 79
bio-energy with carbon capture and sequestration (BECCS) 191, 206, 217
Block Buy Program, US 35, 48, 66–67, 71–75, *74*, 79, 94, 178, 179, 186
booms and busts in solar PV 35–36
Bosch 109, 122
Box, George 212
BP Solar 79, 119, 134, 137, 141
Brightness Program, China 134, 137, 149
BSI 120
Bundesumweltministerium (Environment Ministry), Germany 109, 114
Bündnis 90 110
Bush, George W. 39

Cadmium Telluride (CdTe) 120
California, US 40, 170; LCFS (low carbon fuel standard) 194, 198; Solar Initiative 2006 179; Standard Offer Contracts/ISO4 (Interim Standard Offer Contract #4) 35, 41, 42, 44, 66, 76–77, 108, 112, 161; wind power 31, 41, 77
California Energy Commission (CEC) 76
California Public Utility Commission (CUC) 76–77
Callaghan, W. 73
Calyxo 120
Canada, and China 138
Canadian Solar 149, 152
Carbon Engineering (CE) 194, 198–199, 200

carbon pricing 192–193, 194
Carter, Jimmy 35, 36, 40, 45, 65, 70–71, 75–76, 76–77, 113
CCS (carbon capture and sequestration) 13, 180–181, 192, 206, 207, 217, 218; *see also* DAC (direct air carbon capture and sequestration)
CdTe (Cadmium Telluride) 120
CDU (Christian Democrats), Germany 117
CE (Carbon Engineering) 194, 198–199, 200
CEC (California Energy Commission) 76
cell: definition 46
Centrotherm 107, 123, 124, 126, 153
Chapin, Daryl 58, 59, 60–61
Chernobyl nuclear accident 1986 35, 39, 106, 109
Cherry Hill Conference 65, 67–68
Chichilnisky, Graciela 196
Chile 9, 11
China 2, 17, 24, 31, 47, 133–136, *136*, 169, 170, 178, 179, 187, 218; agriculture 218–219; booms and busts in solar PV 36; Brightness Program 134, 137, 149; company IPOs 134, 148, 155; decarbonization 5; disruptive production 102; domestic market 154–155; early developments 136–139; feed-in tariff 35, 135, 146–147, 155, 180; and Germany 134–135, 138, 144–146, 149, 154; Golden Sun Demonstration Project 155; government support 134, 138–139, 155; international cooperation 137–138; knowledge flows 135, 138; manufacturing equipment 122, 126, 127–128, 143–144, 156, 157, 182; market share in solar PV manufacturing 32, *32*, **33**, *33*; national share of global demand for PV 32, **33**, *33*; New Energy and Development Technology Organization Project 138; NIS (national innovation system) 139, 206; nuclear energy 218; opposition to solar PV 22; "organizational capability" 135, 139, 218; policy windows 39; public policies 46; R&D 45, 135, 136, 137, 149; Renewable Energy Law 2006 35, 42, 135, 138, 147–148; Rooftop Subsidy Program 155; rural electrification program 35, 134, 136, 137, 147; scale of PV 14, *14*, 133; scale up 107, 135–136, 139; silicon supply issues 150–152, *151*; solar industry 148–154, *151*; subsidies 13; Suntech 35, 102, 126, 134, 139–148, 149, 150, 151, 152–153, 157; supply chain 152–153; technology upgrading 155–157, *156*;

turnkey systems 126, 127, 149, 153; wind power 135, 147, 155
China Development Bank 138–139, 149, 155
China Sunenergy 140
Christian Democrats (CDU), Germany 117
CIGS (copper indium gallium) 120
cleanliness of energy supply 2, 3, 4
climate change 183–184, 185, 212, 213; Germany 110; Japan 98; multiple technologies needed for 5–6, 19; reasons for optimism 219–221; as solar PV adoption reason 162; *see also* decarbonization
Climeworks 192, 194–196, 199, 200
CO_2: niche markets for 193–194; *see also* climate change; DAC (direct air carbon capture and sequestration); decarbonization
codified knowledge as innovation accelerator 23, 186, 190; analogous technologies 214; DAC (direct air carbon capture and sequestration) 199, **201**; SNRs (small nuclear reactors) 204, **204**
Colorado, US 10
community shared solar 165
components of solar PV system 46
computer models, errors in 12–13
constant dollars, definition 10
consumer electronics, Japan 86, 87, 93, 95, 96, 97
continuous R&D as innovation accelerator 23, 185, 190; analogous technologies 214; DAC (direct air carbon capture and sequestration) 199, **201**; SNRs (small nuclear reactors) 203, **204**
copper indium gallium (CIGS) 120
cost reductions: DAC (direct air carbon capture and sequestration) 191; installers/installation 160, *160*, *163*, 163–166; local element in 159, 160, *160*, 161, *163*, 163–166, 167
critics of solar PV 13
Crystal Systems 67, 125
crystallized silicon 94, 96, 101, 121, 141, 181
CSG Solar 120
CUC (California Public Utility Commission) 76–77
customer acquisition costs 164

DAC (direct air carbon capture and sequestration) 20, 184–185, *185*, 190–194, *193*, 213; assessment of 199–201, **201**; emerging commercial air capture 194–199; *see also* CCS (carbon capture and sequestration)

Day, R. Evans 56
decarbonization 3–4, 206, 212; government role 6; innovation needs 6–7; multiple technologies needed for 5–6, 19; reasons for optimism 7–8; scale and urgency of 4–5
declining rebates: Germany 114; Japan 99, 100, 108
deforestation 110
demand-pull 16, 17; definition 34; Japan 87; US 66, 71
demographic factors in solar PV adoption 162
Deng Xiaoping 139
Denmark 218
Department of Energy, US *see* DOE (Department of Energy), US
design: dominant 46–47, 181; tolerance for compromises 19
developing countries: energy transitions 21; local financing needs 167, 169
devices 181–182
diffusion furnaces 124, 156
diffusion times 184, 185
direct air capture 169
direct air carbon capture and sequestration *see* DAC (direct air carbon capture and sequestration)
disruptive production 182; analogous technologies 214; China 102; DAC (direct air carbon capture and sequestration) 199, **201**; as innovation accelerator 23, 186, 190, 199, **201**, 204, **204**; SNRs (small nuclear reactors) 204, **204**
DOE (Department of Energy), US 66, 70, 74, 76, 202, 203; "Domestic Policy Review of Solar Energy" 71; National Photovoltaics Program 74
dominant designs 46–47, 181
Dornier 109
DragonTech Ventures 148
drivers of success 177–181
Du Pont 122

early adopters 31, 36, 55
Economics Ministry, Germany 112, 113, 114
economies of scale 37, 42–43, 182, 206; definition 34; as driver of success 179–180; Germany 106; installers 164
Edgar Bronfman Jr. Family 196
Edison, Thomas 56
Edison Electric Institute 168
Edwards, James B. 76
EEG (Erneuerbare-Energien-Gesetz/Renewable Energy Law, 2000, 2004), Germany 35, 39, 40, 41, 42, 106–108, 111, 113–118, *118*, 119, 124, 125, 126

Egypt 10, 169
Einstein, Albert 20, 24, 31, 35, 47, 48, 55, 56–57, 60, 178
Eisenberger, Peter 196
Eisenhower, Dwight 35, 61
electric vehicles 168, 169
Electricity Feed-in Law (Stromeinspeisungsgesetz [StrEG]) 1990, Germany 112
Electrotechnical Laboratory (ETL), Japan 91, 92
Elkem Solar 120
Elysium 203
Energy Conversion Devices 79, 80, 94, 97
Energy Reorganization Act 1974 (US) 69, 70
Energy Research and Development Administration (ERDA) 69, 70, 72, 73, 77–79
energy security 2–3
Energy Technology Innovation Systems see ETIS (Energy Technology Innovation Systems)
energy transitions 111; debates in 21; developing countries 21; EW (Energiewende/Energy Transition) policy, Germany 107, 108, 115, 117
EnergySage 164
enhanced oil recovery, CO_2 use 193–194, 197
Enquette Commission 110, 111
Environment Ministry (Bundesumweltministerium), Germany 109, 114
environmental movement, Germany 106, 109–110
ERDA (Energy Research and Development Administration) 69, 70, 72, 73, 77–79
Erneuerbare-Energien-Gesetz see EEG (Erneuerbare-Energien-Gesetz/Renewable Energy Law, 2000, 2004), Germany
ErSol 119
ETH (Federal Institute of Technology), Switzerland 194
ETIS (Energy Technology Innovation Systems) 15, 16; definition 34
ETL (Electrotechnical Laboratory), Japan 91, 92
EU (European Union): anti-subsidy rules 112, 114; European Commission, Horizon 202 program 195; European Conformity quality standards 144
Eurosolar 110, 113
Eurosolare 145
Evergreen Solar 120
EW (Energiewende/Energy Transition) policy, Germany 107, 108, 115, 117

experience curve 72; see also learning curve
experts' predictions, errors in 11–13, 12
Exxon 21, 62, 79

fabs (fabrication facilities) 34, 42
Federal Energy Administration, US 69
Federal Energy Agency, US 70
Federal Institute of Technology (ETH), Switzerland 194
Federation of Electric Power Companies of Japan 99–100
Federation of German Industry 114
feed-in tariffs: China 35, 135, 146–147, 155, 180; Germany 1–2, 17, 24, 30, 44, 107, 112, 114, 115, 154, 161, 178, 179, 187, 215
Feist, Holger 118
Fell, Hans-Josef 15, 108, 111, 113, 114, 116
Ferrazza, Francesca 145
financial incentives for solar PV adoption 161–162; see also feed-in tariffs; subsidies
Flat Glass 152
Flat Plate Solar Array (FSA) Project 67, 71, 72, 73, 75
food and beverage industry, CO_2 use 193, 195–196, 197
Ford, Gerald 69, 70
Fördverein Solarenergie 110, 113
forest planting 191, 206, 216
formative phases 184, 185
France 5
Fraunhofer-ISE (Fraunhofer Institute for Solar Energy Systems), Germany 109, 186
Free Democrats, Germany 117
Fri, Bob 69, 70
Fritts, Charles 56
FSA (Flat Plate Solar Array) Project 67, 71, 72, 73, 75
Fuji Electric 96
Fuller, Calvin 58, 59

Gao Jifan 149
Gates, Bill 202
Gay, Charlie 67
GCL 152
Gebald, Christoph 194
"general purpose technologies" (GPT) 207, 217
German Physical Society 110
German Solar Energy Industries Association 110
Germany 2, 10, 35, 98, 106–108, 170, 180, 187; booms and busts in solar PV 36; and China 134–135, 138, 144–146, 149, 154; declining rebates 114; early development

108–111; EEG (Erneuerbare-Energien-Gesetz/Renewable Energy Law 2000, 2004) 35, 39, 40, 41, 42, 106–108, 111, 113–118, *118*, 119, 124, 125, 126; EW (Energiewende/Energy Transition) policy 107, 108, 115, 117; feed-in tariffs 1–2, 17, 24, 30, 44, 107, 112, 114, 115, 154, 161, 178, 179, 187, 215; grassroots support for solar PV 22, 106, 108–109; manufacturing equipment 93, 107, 108, 115, 123–129, *126*, **128**, 179; market share in solar PV manufacturing 32, *32*, **33**, *33*; national share of global demand for PV 32, **33**, *33*; NIS (national innovation system) 108; nuclear energy 115, 117; public policies 46; R&D 80, 107, 110; rooftop programs, 1988–1999 111–113, 116; scale of PV 14, *14*, 183; SPD-Green coalition 1998 39, 113; subsidies 13, 48; wind power 31, 112, 116; *see also* Q-Cells
Gillis, J. 107
Gintech 153
Global Environmental Facility 137
global financial crisis 2008 39, 121, 153, 155
"global innovation systems" 44–45
global mobility, as innovation accelerator 23, 187, 190; analogous technologies 215; DAC (direct air carbon capture and sequestration) 200, **201**; SNRs (small nuclear reactors) 204, **204**
Global Solar Fund (GSF) 147, 153
Global Thermostat (GT) 194, 196–198, 199, 200
globalization, and knowledge flows 44
Goetzberger, Adolf 109
Goldman Sachs 148
Good Energies 119
GP Solar 123–124
GPT ("general purpose technologies") 207, 217
Green, Martin 15, 39, 44, 78, 80, 133, 139, 140–141, 143, 144, 187
green identity, as solar PV adoption reason 162
Green Party, Germany 110, 113, 114, 117, 154
Green Revolution 206, 216
greenhouse cultivation, CO_2 use 193, 195, 197
Greenpeace 12, 110, 113
Grunow, Paul 118
GSF (Global Solar Fund) 147, 153
GT (Global Thermostat) 194, 196–198, 199, 200
GT Advanced technologies 151
GT Solar 151
GTAT 123, 149, 151

Hallwachs, Wilhelm 56
Hanwha Solar 121
hardware costs: definition 10, 46
Hayakawa, Tokuji 15, 88–89
Hayes, Dennis 76
HCT Cutting Technologies 123, 125
HCT Shaping Systems 78
Hemlock 102, 124, 151
Hertz, Heinrich 56
Hitachi 93, 96
Hoffman, Les 36, 62
Hoffman Electronics 62, 89
Hogg, David 133, 140–141
Holos Generators 203
Horigome, Takashi 91, 94
hybrid solar cells 179, 181
Hydrocell 194
hydrogen 39

IAEA 215
Iceland 195
Idaho National Laboratory 202
IEA (International Energy Agency) 98, 115–116, 215
India 10, 128, 140, 162, 169; scale of PV 14, *14*
individuals, in knowledge transfer 43
Innolas 123
innovation: definition 34; diffusion times 184, 185; formative phases 184, 185; need for 6–7; *see also* NIS (national innovation systems)
innovation accelerators 185–188; analogous technologies 20, 185–188, 212–215; codified knowledge 23, 186, 190, 199, **201**, 204, **204**, 214; continuous R&D 23, 185, 190, 199, **201**, 203, **204**, 214; disruptive production 23, 186, 190, 199, **201**, 204, **204**; global mobility 23, 187, 190, 200, **201**, 204, **204**, 215; political economy 23, 187, 190, 200, **201**, **204**, 205; public procurement 23, 185–186, 199, **201**, 203, **204**; robust markets 23, 186–187, 190, 200, **201**, 204, **204**, 214–215; trained workforce 23, 186, 190, 199, **201**, 203–204, **204**, 214
innovation systems 16; international 44–45; *see also* ETIS (Energy Technology Innovation Systems)
installers/installation 31, 40; competition 164; cost reductions 160, *160*, *163*, 163–166; customer acquisition costs 164; economies of scale 164; Germany 107; installed cost, definition 10; Japan 98; knowledge spillovers 166–167, 170;

learning by 160, 163–164, 170; permit and inspection system 165
Intergovernmental Panel on Climate Change 116
intermittance 167–168
international cooperation on climate change 220
International Electrotechnical Commission certification 144
International Energy Agency (IEA) 98, 115–116, 215
international innovation systems 44–45
International Renewable Energy Agency (IRENA) 116
International Standards Organization 9001 certification 144
inverter: definition 46
Iranian Revolution and oil crisis 1979 39, 70, 94, 109, 178
IRENA (International Renewable Energy Agency) 116
ISO4 (Interim Standard Offer Contract #4)/Standard Offer Contracts, California, US 35, 41, 42, 44, 66, 76–77, 108, 112, 161
Italy 107, 195

JA Solar 138, 140, 141
Japan 2, 85–87, 161, 162, 170, 187; declining rebates 99, 100, 108; early developments 88–89; knowledge flows 87; lull, 1984–94 96–98; market share in solar PV manufacturing 31, 32, *32*, **33**, *33*; national share of global demand for PV 32, **33**, *33*; New Energy and Development Technology Organization Project 138; New Sunshine Program 35, 44, 99, 100; niche markets 17, 30, 62, 86, 87, 101, 178; NIS (national innovation system) 86–87; nuclear energy 90, 92, 99–100; opposition to solar PV 22; public policies 46; R&D 36, 45, 85–86, 86–87, 90, 92, 93, 94–95, *95*, 96–97, 99; Rooftop Subsidy Program 1994–2005 35, 86, 87, 97, 98–100, *100*, 103, 179, 180; scale of PV 14, *14*; Sharp's scale-up 100–103; solar PV funding *95*; subsidies 13, 17, 24, 30; Sunshine Project, 1974–1984 35, 87, 90–96, 97, 98
Japan Photovoltaic Energy Association (JPEA) 97, 98
Japan Solar Energy Company (JSEC) 96
Japanese National Aeronautics and Space Development Agency 89
Japanese Scientific Satellite Program 89
Jet Propulsion Lab (JPL) 70, 71, 72

Jiajun Wang 142
Jianhua Zhao 140
Jinko 153, 154, 192
Jiqun Shi 140
Jonas & Redmann 123, 124
JPEA (Japan Photovoltaic Energy Association) 97, 98
JPL (Jet Propulsion Lab) 70, 71, 72
JSEC (Japan Solar Energy Company) 96

Kaifeng Solar Cell Factory 137
Kashiwagi, T. 90
Keith, David 198
Kelly, Mervin 58
Ken, Suzuki 85, 91, 92
Kenjiro, Kimura 93
knowledge *see* codified knowledge as innovation accelerator
knowledge flows 32, 33, 43–45, 66, 179; China 135, 138; Japan 87
knowledge spillovers: analogous technologies 215; DAC (direct air carbon capture and sequestration) 200, **201**; as driver of success 179, 204, **204**; as innovation accelerator 23, 187, 190, 200, **201**, 204, **204**; installers/installation 166–167, 170; SNRs (small nuclear reactors) 204, **204**
Kohl, H. 110
Korea 126; NIS (national innovation system) 206
Kotenko, Dmitry 151
Kressin, Manfred 118
KWh: definition 9
Kyocera 86, 87, 92, 96, 97, 100, 123, 125
Kyoto Protocol 1997 5, 98, 220

Laird, Frank 69
landline telephone industry 168
Law for the Development and Promotion of Oil-substitute Energy (Law No. 71), Japan 94
LbD (learning by doing) 163–164, 170, 179
LCFS (low carbon fuel standard), California, US 194, 198
LDK 120, 151–152
LDK Solar 138
leaded gasoline 220
leadership in solar PV 31–36, *32*, **33**, *33*, 40
leapfrogging 22–23
learning curve 48, 115, 117; *see also* experience curve
leased systems 164–165
Leibniz Association 109
Lemoine, Rainer 118

Lenard, Philip 55
Liansheng Miao 149
Liberia 11
lighthouses 89
Limits to Growth studies, Club of Rome 109–110, 113
Linde 195
List, Friedrich 15
Liu Lifeng 156–157
local element in solar PV 40–41, 159–161, *160*, 170; adoption processes 159, 161–163; challenges 167–169; cost reductions 159, 160, *160*, 161, *163*, 163–166, 167; knowledge spillovers 166–167, 170; opportunities 169–170
lock-in 23
Lovins, Amory 111

Malaysia 40
Mali 62
mangrove planting 191
Manhattan Project 66, 68
manufacturing equipment 125, 126, *126*, **128**, 182, 187; China 122, 126, 127–128, 143–144, 156, 157, 182; Germany 93, 107, 108, 115, 122, 123–129, *126*, **128**, 179; and Q-Cells 122; R&D 125, 126, *126*; turnkey systems 107, 126–128, 149, 153, 169; *see also* disruptive production
manufacturing of solar PV **33**, *33*; global market shares 31–32, *32*
Manz 123
Margolis, R. 164
market failures 6, 167
Masato, Nebashi 85
Matsumoto, M. 90
Matsushita 93, 96
Maycock, Paul 15, 48, 69, 70, 72, 74, 76
McDonald, R. 73
McKinsey 119
Meier Brothers 194
MEMC (Monsanto Electronic Materials Company) 124, 150, 151
METI (Ministry of Economy, Trade, and Industry), Japan 88
Mexico 9
Meyer-Burger 123, 125, 148
micro-algae cultivation, CO_2 use 193, 197
micro-reactors *see* SNRs (small nuclear reactors)
Middle East 169
Millikan, Robert 55, 57
Milner, Anton 118–119, 121
Minchin, George 56

Ministry for International Trade and Industry (MITI), Japan 85, 86–87, 88, 89, 90, 96, 97, 98, 99
Ministry of Economy, Trade, and Industry (METI), Japan 88
Mission Innovation 214, 215
MITI (Ministry for International Trade and Industry), Japan 85, 86–87, 88, 89, 90, 96, 97, 98, 99
Mitsubishi 89, 151
Mitsubishi Electric 96
Mobil 96
modular scale 19, 30; as driver of success 179–180
module: definition 46
mono-crystalline silicon 89, 125, 137, 145
Monsanto Electronic Materials Company (MEMC) 124, 150, 151
Moore's Law 46
MSK 148
multi-crystalline technology 78
MWh: definition 9

National Development and Reform Commission (NDRC), China 147
National Electrotechnical Laboratory, Japan 94
National Fabricated Products (NFP) 62
national innovation systems *see* NIS (national innovation systems)
National Renewable Energy Laboratory (NREL), US 66, 111, 186
National Science Foundation (NSF), US 66, 67
natural gas: PPA (Power Purchase Agreement) 10; US 13
navigation buoys 89, 180
NDRC (National Development and Reform Commission), China 147
NEC 89, 93, 96
NEDO (New Energy Development Organization), Japan 92, 93, 94, 95, 96, 97, 99, 186
negative emissions technologies (NETs) 19, 190–191, 192
Neosolar Power 153
net metering 98, 111, 161
Netherlands 138
NETs (negative emissions technologies) 19, 190–191, 192
Nevada, US 9
New Energy Development Organization (NEDO), Japan 92, 93, 94, 95, 96, 97, 99, 186

New Sunshine Program, Japan 35, 44, 99, 100
NFP (National Fabricated Products) 62
NGOs (non-governmental organizations) 110
niche markets 1, 19, 30, 36–38, *38*, 55, 62, 159, 162–163, 183, 186–187; DAC (direct air carbon capture and sequestration) 193–194; definition 34; as driver of success 180; Japan 17, 30, 62, 86, 87, 101, 178; SNRs (small nuclear reactors) 201–202
Ningbo Solar Power Source Factory 137
NIS (national innovation systems) 15–16, 45, 206; China 139, 206; definition 34; Germany 108; Japan 86–87, 139; Korea 206; and leadership in solar PV 32–34; and leapfrogging 22–23; right innovation models 217–219; US 66, 218
Nitol 150–151
Nixon, Richard 35, 65, 66, 67, 68–69, 76, 90
non-governmental organizations (NGOs) 110
NPC 145
NRC (Nuclear Regulatory Commission), US 202
NREL (National Renewable Energy Laboratory), US 66, 111, 186
NSF (National Science Foundation), US 66, 67
nuclear energy 1, 13, 21, 61–62, 180–181, 206–207, 218; Chernobyl nuclear accident 1986 35, 39, 106, 109; China 218; competition with solar PV 22; and decarbonization 5; Germany 115, 117; Japan 90, 92, 99–100; US 69, 187; *see also* SNRs (small nuclear reactors)
Nuclear Regulatory Commission (NRC), US 202
Nukem/Schott 124
NuScale 202

OAPEC (Organization of Arab Petroleum Exporting Countries) 67
observation, in knowledge transfer 43–44
OCI 152
Oerlikon 151
Office of Alternative Energy Policy, Japan 91
Ohl, Russel Shoemaker 58
oil companies: early adopters of solar PV 55, 62, 63, 79.80; and globalization 44
oil crises 24, 35, 70; Arab Oil Embargo 1973 3, 39, 65, 67–69, 79, 90, 91, 109, 113, 178; Iranian Revolution and oil crisis 1979 39, 70, 94, 109, 178

oil price collapse, 1985 35
Oklo 203
Öko Institute 110, 111
"Operation Lunch Box" 61
Operation Paperclip 61
optimism, reasons for 7–8, 219–221
"Ordo-liberalism" 110
organic solar cells 47
Organization of Arab Petroleum Exporting Countries (OAPEC) 67
O'Shaunessy, E. 164
Ovshinky, Stan 94
ozone layer, hole in 220

p-n junction 47, 57, 58, 59–60, 89, 96, 124, 178, 181
PACE (Property Assessed Clean Energy) 165
Pacific Power 133
Pacific Solar 133, 140–141, 143, 145
Paley Commission 1953 61, 67
Paris Agreement on Climate Change 2015 4, 107, 184, 191, 220
Pauling, Peter 148
PCAST (President's Council of Advisors on Science and Technology) 6–7
Pearson, Gerald 58, 59
PECVD (plasma enhanced chemical vapor deposition) 125, 146
peer effects in adoption decisions 160, 162, 183
Peng Xiaofeng 151
permit and inspection system for installers 165
perovskite solar cells 47, 179, 181
Persian Gulf 10
Philips 62, 109
photoelectric effect 18, 20, 35, 46, 47, 48, 55, 56, 57, 58, 178
Photowatt 149
plasma enhanced chemical vapor deposition (PECVD) 125, 146
Poland 5
policy stream in policymaking 39
policy support, as driver of success 180
policy windows 35–36, 39
political advocacy 160–161
political economy 22; analogous technologies 215; DAC (direct air carbon capture and sequestration) 200, **201**; Germany 108; as innovation accelerator 23, 187, 190, 200, **201**, **204**, 205; SNRs (small nuclear reactors) **204**, 205
politics stream in policymaking 39
poly-crystalline silicon 145, 151

PPA (Power Purchase Agreement): definition 10; natural gas 10; price of 37, *38*
preserved knowledge 45
President's Council of Advisors on Science and Technology (PCAST) 6–7
Preussen Elektra 112
Prince, Morton 59, 62, 74
problem stream in policymaking 39
Project Independence, US 35, 65, 66, 67, 68–69, 71, 90
Project Megawatt, Switzerland 98, 111
Property Assessed Clean Energy (PACE) 165
public policy and solar PV 38–43
public procurement 24; analogous technologies 214; DAC (direct air carbon capture and sequestration) 199, **201**; as innovation accelerator 23, 185–186, 199, **201**, 203, **204**; SNRs (small nuclear reactors) 203, **204**
PURPA (Public Utilities Regulatory Policy Act) 1978 76–77, 112
PV (photovoltaics): definition 46; *see also* solar PV
PV Interim Act 2011, Germany 117
PV Power Generation Technology Research Association (PVTEC), Japan 97
PV Science and Engineering Conference (PVSEC) 92
PV system: definition 46
PVA Tepla 125
PVSEC (PV Science and Engineering Conference) 92
PVTEC (PV Power Generation Technology Research Association), Japan 97

Q-Cells 102, 106, 107, 118–123, 126, 141, 146, 192
quality of life 221
quantum dot solar cells 47, 179

racking: definition 46
RANN (Research Applied to National Needs) program, NSF (National Science Foundation), US 67
Rappaport, Paul 70
Raymond, Lee 21
RCA 94
R&D 36; China 45, 135, 136, 137, 149; DAC (direct air carbon capture and sequestration) 191; Germany 80, 107, 110; Japan 36, 45, 85–86, 86–87, 90, 92, 93, 94–95, *95*, 96–97, 99; US 17, 30, 36, 45, 65–71

Reagan, Ronald 36, 40, 45, 66–77, 71, 74, 75–76, 80, 86, 113
Real, Markus 44, 78, 80, 111
REC 120, 151
reforestation 191, 206, 216
reliability of energy supply 2–3, 4
Rena 107, 123
Renesola 152
Renewable Energy Law 2006, China 35, 42, 135, 138, 147–148
Renewable Energy Sources Acts 2009, 2012, 2014, Germany 117
Renova 151
retail electricity prices: definition 9
ribbon silicon 96, 123
robust markets, as innovation accelerator 23, 186–187, 190; analogous technologies 214–215; DAC (direct air carbon capture and sequestration) 200, **201**; SNRs (small nuclear reactors) 204, **204**
Roessner, J. D. 74
Roth & Rau 107, 123, 146
rural electrification program, China 35, 134, 136, 137, 147
Russia 5
Rwanda 11
RWE 109
RWE-Schott 119

S-400 solar cell 62
SAMICS (Solar Array Manufacturing Industry Costing Standards) 73
Sandia National Laboratory 70
Sanyo 86, 87, 89, 92, 95, 97, 100, 137
Sanyo Electric 96
satellites 89, 134, 180; Sputnik launch 1957 39, 55, 62, 178
Saudi Arabia 9, 169–170
scale up 179, 192; analogous technologies 213–215; China 107, 135–136, 139; DAC (direct air carbon capture and sequestration) 192–193, 200; definition 34
Scheer, Hermann 15, 108, 110, 111, 113, 114, 116, 118
Schlesinger, James R. 70
Schmid 107, 123, 126
Schott Solar 118
scientific origins and understanding 17–18, 19, 24, 30, 47, 55–61, *60*, 178, 181; emerging commercial market 61–63; key scientists 56–57
Seamans, Robert C. 69, 70
selenium PV cells 56, 59
self-sufficiency, as solar PV adoption reason 162
September 11, 2001 39

SERI (Solar Energy Research Institute), US 66, 70, 76, 77–79, 111
Sharp 36, 62, 79, 86, 88–89, 92, 93, 97, 125, 136, 137, 141, 144, 148; scale up 87, 100–103, 192
Shell 79, 112
Shell Solar 119, 134, 137
Shi Zhengrong 15, 133, 139–147, 152, 154
Shimamoto, M. 91
Siemens (company) 55, 57, 78, 79, 80, 111, 125
Siemens, Werner von 56
Siemens Solar 137
silicon 46–47, 58, 67; a-Si (amorphous silicon) 94, 95–96, 97, 103; cost and supply issues 102–103, 117, 119–120, 121, 124, 150–152, *151*, 152; mono-crystalline silicon 89, 125, 137, 145; poly-crystalline silicon 145, 151; ribbon silicon 96, 123; thin film silicon 87, 101, 103, 120, 126, 140, 141, 181, 186
Six-Day War 1967 67
Slovakia 5
smart grid technology 1
Smee, Alfred 56
Smith, Willoughby 56
SMRs (small modular reactors) *see* SNRs (small nuclear reactors)
SNRs (small nuclear reactors) 20, 190, 201–203; assessment of 203–205, **204**; *see also* nuclear energy
soft costs 8–9, 17; cost reductions by installers 160, *160*, *163*, 163–166; definition 10; local element in 40–41
"soft energy paths" 115
soil, carbon retention in 191, 206, 216
Solar Array Manufacturing Industry Costing Standards (SAMICS) 73
Solar Energy Research Institute (SERI), US 66, 70, 76, 77–79, 111
Solar Initiative 2006, California, US 179
Solar Power Corporation 62, 79.80
solar PV: characteristics of 19, 181–183; competition with nuclear energy 22; costs of 1, 8–11; future potential of 13–14, *14*; list of applications for 37; milestones in history of 35; national demand for 32, **33**, *33*; opposition to 22; overview of development 16–18, *18*; production process 182; public perceptions of 11; research questions 14–15; storage options 168; success of 8–18, *12*, *14*, *18*; system components 46; technical progress in 46–48; theoretical basis for 47

solar PV model 18–23, 177; acceleration of 20–21, 23, 185–188; applications of 190–207; drivers of success 177–181; PV characteristics 181–183; slowness of 183–185, *185*
Solar PV R&D Act 1978 (US) 70
Solar Technology International (STI) 79
solar thermal electric (STE) 77, 92
solar vehicle batteries 149
Solarex 79, 123
Solarfun 153
Solartech 153
Solarworld 79, 107, 119, 124, 125, 126, 146
Solibro 120
Solon AG 118
SongDian Dao Xiang (Township Electrification Program), China 137
Sonto GmbH 120
South Africa 11
South Korea 162
Southeast Asia 169
space programs 36–37, 55, 62, 89, 136, 178
Spain 40, 107, 149, 153, 180, 187
SPD, Germany 39, 113, 114, 117
Special Account for Alternative Energy Development, Japan 94, 96, 99
Spectrolab 141
Spire Corp 79, 80
Sputnik launch 1957 39, 55, 62, 178
SRI International 196
Standard Offer Contracts/ISO4 (Interim Standard Offer Contract #4), California, US 35, 41, 42, 44, 66, 76–77, 108, 112, 161
standardized products 181–182
standards and certification, China 135, 144, 148
Star-Core Nuclear 203
State Grid 155
STE (solar thermal electric) 77, 92
Steinberger, Markus 107
STI (Solar Technology International) 79
Stoher, Immo 118
storage options for solar PV 168
Stromeinspeisungsgesetz [StrEG] (Electricity Feed-in Law) 1990, Germany 112
subsidies 13, 24, 165–166, 167; Germany 13, 48; Japan 13, 17, 24, 30; *see also* feed-in tariffs
success, drivers of 177–181
Sumitomo 151
Sunergy 153
SunPower 79–80
Sunshine Project, 1974–1984, Japan 35, 87, 90–96, 97, 98

Suntech 35, 102, 126, 134, 139, 148–149, 150, 151, 152, 157, 192; collapse of 154; initial business plan 141–143; IPO (initial public offering) 148; Line #1: start up 143–144; Line #2: Germany 144–146; Line #3: feed-in tariff 146–147; and Pacific Solar 140–141; and the Renewable Energy Law 2006 147–148; supply chain 152–153; and UNSW 139–141
Sunways 118, 126
Swanson, Dick 79, 80
Sweden 5
Switzerland 78, 80, 194
system failures 6
system integration challenges 180–181
Szpitalak, Ted 141, 143

Taiwan 126
Tanzania 10–11
technology lock-in 23
technology-push 16, 17, 87; definition 34
technology roadmaps, Japan 86
technology spillovers 19, 31, 182; Japan 87, 96, 97
technology types 205–207, **207**; right innovation model for 215–217, *216*
Tempress/Amtech 123
TerraPower 202
thin film silicon 87, 101, 103, 120, 126, 140, 141, 181, 186
third-party ownership (TPO) systems 164–165
"three streams" in policymaking 39
TI (Texas Instruments) 48, 69, 70, 72
Tideland Signal 62
Tokuyama 102
Toshiba 93, 96
Township Electrification Program (SongDian Dao Xiang), China 137
Toyo Silicon 93
TPO (third-party ownership) systems 164–165
trained workforce, as innovation accelerator 23, 186, 190, 199; analogous technologies 214; DAC (direct air carbon capture and sequestration) 199, **201**; SNRs (small nuclear reactors) 203–204, **204**
Transatomic 202
"transnational linkages" 45
Traveling Wave Reactor (TWR) 202
trichlorosilane 151
Trina 147, 149, 152, 154
Truman, H. 36

Turkey 128, 169
turnkey systems 107, 126–128, 169; China 126, 127, 149, 153
TWR (Traveling Wave Reactor) 202
Tyco 93, 123

U-Battery 203
Ultra-Safe Nuclear Corporation 203
UN Copenhagen Climate Change Conference 2009 7, 219
UN Scientific Conference on the Conservation and Utilization of Resources 1949 61
UNSW (University of New South Wales) 77, 133, 139–140, 142, 144, 150, 152, 154, 178–179, 186
US 2, 3; 45-Q tax credits 194, 198; agriculture 218; Block Buy Program 35, 48, 66–67, 71–75, *74*, 79, 94, 178, 179, 186; booms and busts in solar PV 36; emergence of PV industry 79–80; employment in solar PV 78; financial incentives for solar PV adoption 161; market share in solar PV manufacturing 31, 32, *32*, **33**, *33*; military adoption of solar PV 24, 36, 55, 61, 63; national share of global demand for PV 32, **33**, *33*; natural gas 13; NIS (national innovation system) 66, 218; nuclear energy 69, 187; opposition to solar PV 22; Project Independence 35, 65, 66, 67, 68–69, 71, 90; PTC (Production Tax Credit) 42, 117; public policies 46, 65–76, 77–79; public procurement policies, 1970s 17, 30; Public Utilities Regulatory Policy Act 1978 41; R&D 17, 30, 36, 45, 65–71, 86, 178; RPS (renewable portfolio standards) policies 42; scale of PV 14, *14*; soft costs 160, *160*, 163, *163*; technology-push approach 65–71, *68*
US Army Signal Corps 44, 61
US Coast Guard 62
US Navy satellite programs 62
utility companies, threat of solar PV to 167, 168

vehicle batteries, solar 149
vehicle fuel economy 1
vehicle-to-grid service companies 169
Von Fabeck, Wolf 111

Wacker 102, 109, 124, 125, 151
Warren-Alquist Act 1974, California 76–77
Wene, Clas-Otto 115

Wenham, Stuart 152
Western Initiative for Nuclear (WIN) program 202
Western Solar Energy project 138
Westinghouse 202
wholesale electricity prices: definition 9
willingness-to-pay (WTP) 162–163
WIN (Western Initiative for Nuclear) program 202
wind power 17, 31, 41–42; California, US 31, 41, 77; China 135, 147, 155; Germany 31, 112, 116
Wolfe, Phillip 79
workforce *see* trained workforce, as innovation accelerator
World Bank 137

WTP (willingness-to-pay) 162–163
Wurzbacher, Jan 194

X-Energy 202
Xiaohua Qu 149
Ximin Dai 140

Yang, Huaijin "Sammy" 141, 142
Yerkes, Bill 79
Yingli 138, 147, 149, 151, 152, 154
Yokuyama 151
Yom Kippur War 1973 67
Yunnan Semiconductor Devices factory 137
Yuting Wang 149

Zeigler, Hans 61